Experienc

Voices from the Bottom

D. Stanley Eitzen
Colorado State University

Kelly Eitzen Smith
University of Arizona

Australia • Canada • Mexico
Singapore • Spain • United Kingdom
United States

THOMSON
WADSWORTH

Senior Executive Editor, Sociology: *Sabra Horne*
Assistant Editor: *Stephanie Monzon*
Editorial Assistant: *Matthew Goldsmith*
Technology Project Manager: *Dee Dee Zobian*
Marketing Manager: *Matthew Wright*
Marketing Assistant: *Michael Silverstein*
Signing Representative: *Jane Pohlenz*
Project Manager, Editorial Production: *Emily Smith*
Print/Media Buyer: *Barbara Britton*
Permissions Editor: *Bob Kauser*
Production Service: *Ruth Cottrell*
Copy Editor: *Ruth Cottrell*
Cover Designer: *Preston Thomas*
Compositor: *TBH Typecast, Inc.*
Text and Cover Printer: *Webcom Limited*

Printed in Canada

1 2 3 4 5 6 7 06 05 04 03 02

For more information about our products, contact us at:
Thomson Learning Academic Resource Center
1-800-423-0563

For permission to use material from this text, contact us by:
Phone: 1-800-730-2214
Fax: 1-800-730-2215
Web: http://www.thomsonrights.com

Library of Congress Control Number: 2002106306

ISBN 0-534-52958-5

Wadsworth/Thomson Learning
10 Davis Drive
Belmont, CA 94002-3098
USA

Asia
Thomson Learning
5 Shenton Way #01-01
UIC Building
Singapore 068808

Australia
Nelson Thomson Learning
102 Dodds Street
South Melbourne, Victoria 3205
Australia

Canada
Nelson Thomson Learning
1120 Birchmount Road
Toronto, Ontario M1K 5G4
Canada

Europe/Middle East/Africa
Thomson Learning
High Holborn House
50/51 Bedford Row
London WC1R 4LR
United Kingdom

Latin America
Thomson Learning
Seneca, 53
Colonia Polanco
11560 Mexico D.F.
Mexico

Spain
Paraninfo Thomson Learning
Calle/Magallanes, 25
28015 Madrid, Spain

If statistics are the language of the "New Economy," the personal story is the language of the "Under Economy."

Beth Kelly, Director of People United for Families (2001)

Many in our country do not know the pain of poverty. But we can listen to those who do.

George W. Bush (Inaugural Address, 2001)

About the Editors

D. Stanley Eitzen (Ph.D., University of Kansas, 1968) is professor emeritus in sociology from Colorado State University. He is author or co-author of seventeen books, many of which are in multiple editions. His books that fit with this book are *Social Problems,* 8th edition (Allyn and Bacon, 2000); *Solutions to Social Problems: Lessons from Other Societies,* 2nd edition (Allyn and Bacon, 2001); *Paths to Homelessness: Extreme Poverty and the Urban Housing Crisis* (Westview, 1994); and *The Reshaping of America* (Prentice Hall, 1989). His major research interests are social problems, social inequality (race/class/gender/sexuality), families, power, and sport.

Kelly Eitzen Smith (Ph.D., University of Arizona, 1999) is adjunct professor of sociology and assistant director of the Center for Applied Sociology at the University of Arizona. Her dissertation, under the direction of David A. Snow, was an ethnography of homeless women. While a graduate student she was part of a large research project where she interviewed the homeless in Detroit, Philadelphia, and Tucson. She has done extensive volunteer work in homeless shelters in Tucson. Her sociological interests are gender, social inequality, and social movements.

Contents

Preface

Sociologists, political scientists, and economists have counted and classified the poor by income, race, ethnicity, gender, household and family status, and geographical location. These social scientists have also described how the poor live and theorized as to why they are poor. There are important studies that look at poverty from the bottom up (e.g., Kozol, 1988; Liebow, 1993; MacLeod, 1995; and Snow and Anderson, 1993), but many of them have provided disembodied information that ignores what it is like to be poor from the perspective of the poor themselves. Our goal is to add to their insights about poverty by focusing on the voices of the poor, not just as they are filtered by social scientists. We seek to know how the poor define their situations; the reasons for their plight; the problems they face; their coping strategies for providing the daily necessities of food, clothing, shelter, transportation, and health care; and their perceptions of government and charity efforts.

If we ignore the voices of the poor, we have dehumanized them, making their humanity invisible. As Eric Alterman argued in an essay for *The Nation* (2000:12), "While we hear that 43 million Americans lack health insurance, rarely do we read or hear about what it's like to raise a sick child without it." Similarly, we know little about how working-poor families manage when they pay more than half of their income for housing; how the shift from a manufacturing economy to one based on knowledge/service has left them behind; how the 1996 welfare legislation has affected former AFDC mothers who now work for minimum wages while paying for child care, transportation, clothing, and other costs of working outside the home; how the poor pay more for goods and services. Hidden in their answers to these and other questions is the answer to the core question of whether the reason for poverty is fundamentally one of character or is the result of structural forces.

Without hearing the voices of the impoverished our understanding of their social life is incomplete. Analyses from the top down miss the insights that only those experiencing poverty can articulate. This omission has important implications for those policymakers interested in solving social problems. In this regard, Alan Wolfe (1991:23) has said:

> If policymakers are going to respond adequately to the transformations America has been experiencing, they ought first to consult, not quick and shallow public opinion polls nor the well-financed view of those with an immediate stake in whatever policy is being debated, *but instead those whose lives constitute and are constituted by the policies they make.* It is a long way from the "bottom up" to the "top down"—longer than from the "top down"

> to the "bottom up"—but like many longer journeys, the rewards at the end of the trip are more lasting. (emphasis added)

In addition to the descriptions of the poor that ignore their voices, the poor are mostly invisible to the middle and upper classes (the primary social locations of most college students) who live their lives apart from the poor and the working class. As Barbara Ehrenreich has noted (2000:19):

> These fortunates inhabit an increasingly insular world of their own, far from the customary venues of the poor or even the working class. They live in fortress-like apartment buildings, gated communities, or inaccessible exurbs. They do not use public transportation and are unlikely to send their children to public schools.

This collection of readings will provide the voice, the presence, and the perspective of the poor who live on the margins and are generally invisible to the rest of us. Our goals are twofold: (1) to bring the realities of the lives of the impoverished as close to the reader as possible; and (2) to get the reader to listen carefully to these voices of the poor in order to enhance their understanding of:

- How the poor became poor.
- How the poor are treated by individuals and organizations in the community.
- What keeps the poor poor.
- How the poor manage day-to-day.
- What theory of causation best explains poverty.
- The consequences of the welfare-to-work federal legislation.
- The best solution for ending poverty.

We seek answers to these issues by listening to those on the bottom. To try to accomplish this, we divided the book into five parts. Part I provides an overview of poverty in the United States and a sociological perspective for understanding this social problem. Part II, using voices of the poor, seeks answers to the question: Why are the poor poor? Parts III and IV explore the impact that society's institutions have on the lives of poor people. The final section examines agency efforts by poor people to create viable lives even when they face difficult situations. These efforts include the organization and use of an underground (hidden, irregular, and sometimes illegal) economy; family networks for exchange of goods and services; neighborhood organizations; and collective efforts such as strikes and boycotts to change work conditions, welfare regulations, and school policies.

We have selected a variety of essays that provide the voices of the poor. The voices come from three sources: (1) from those who are currently living in poverty; (2) from those who were once poor and look back on their experiences while impoverished; and (3) from social scientists who include voices of the poor along with their analyses.

To aid the reader we provide an introductory essay that puts poverty in perspective by providing data on the extent and distribution of poverty in the United States. It concludes with a discussion of the sociological imagination as a way of seeing and understanding poverty through the experiences of those at the bottom of society. The remaining four parts and the chapters within each one are preceded by a brief overview that provides the context and alerts the reader to the important issues discussed. Each essay opens with an overview that contextualizes the essay and provides focus questions to direct the reader's attention to key points in the article. Each part concludes with questions that link the articles together. Pertinent Web sites, scholarly references, and a glossary are also included.

ACKNOWLEDGMENTS

We are indebted to the following reviewers whose critiques were unusually helpful: Leon Anderson, Ohio University; Marino Bruce, University of Wisconsin, Madison; Doug Harper, Duquesne University; and Lynda D. Nyce, Bluffton College.

Several friends and colleagues have helped us by providing ideas and essays that we have included. Our very special thanks in this regard to Kathryn D. Talley, Michelle Navarre Cleary, and Beth Kelly. In a more general way we are indebted to Betty Eppler of Casa Paloma in Tucson, the staff at Las Amigas in Tucson, David A. Snow, and Maxine Baca Zinn. Special thanks to Eve Howard, our editor, for her support of this project and for her suggestion and implementation of a grant to two agencies in Tucson that serve women in poverty.

D. Stanley Eitzen and Kelly Eitzen Smith

PREFACE BIBLIOGRAPHY

Alterman, Eric (2000). "Still With Us," *The Nation* (April 24):12.

Ehrenreich, Barbara (2000). "The Vision-Impaired Rich," *The Progressive* 64 (May 2000): 18–19.

Kozol, Jonathan (1988). *Rachel and Her Children: Homeless Families in America* (New York: Crown).

Liebow, Elliot (1993). *Tell Them Who I Am: The Lives of Homeless Women* (New York: The Free Press).

MacLeod, Jay (1995). *Ain't No Makin' It: Aspirations and Attainment in a Low-Income Neighborhood* (Boulder, CO: Westview Press).

Snow, David A., and Leon Anderson (1993). *Down on Their Luck: A Study of Homeless Street People* (Berkeley: University of California Press).

Wolfe, Alan (1991). "Introduction." Pp. 1–13 in *America at Century's End,* Alan Wolfe (ed.). (Berkeley: University of California Press).

PART I

Poverty in the United States and the Sociological Imagination

This introductory essay provides an overview in two ways. First, as background, it provides the official statistics on poverty in the United States in terms of extent and distribution. While necessary and important, these numbing statistics miss what is happening in individual lives, a void that the essays in this book will fill. Second, hearing and understanding these voices requires a way of listening and interpreting sociologically—"sociological imagination"—a perspective that is elaborated in the final section of this essay.

THE EXTENT AND DISTRIBUTION OF POVERTY IN THE UNITED STATES

A common boast by Americans, and one not without some truth, is that the United States is the richest country on earth. The United States has great natural resources, the most advanced technology, and a very high standard of living. At the same time, when compared with other industrial democracies of the West, the United States has the highest proportion of its population living in poverty (Smeeding & Gottschalk, 1998). Consider the following international comparisons, where the United States does not fare well:

- U.S. Women Connect compiled a report card on the U.S. government's efforts to improve equality for women and gave it an "F" for its attempts to reduce poverty among American women (reported in Winfield, 2000).

- Among the industrialized nations, the United States has the highest rate of child poverty (one in six). Nearly 11 million children (15.5 percent of all children) have no health insurance, and 500,000 children are homeless. The infant mortality rate in some U.S. inner cities rivals that of Malaysia (Pollitt, 2001).
- Compared with Canada and the nations of Western Europe and Scandinavia, the United States eliminates much less poverty than any of the other 14 nations through welfare subsidies (Solow, 2000).

In effect, then, despite its richness, the United States has a serious problem with poverty.

This section provides a brief survey of what is known about poverty in the United States. It furnishes facts about the impoverished—how poverty is defined and how poverty is distributed by race, gender, age, and place.

The Poverty Threshold

This official poverty line (**poverty threshold**) is based on what the Social Security Administration considers the minimal amount of money required for a subsistence level of life. The number is computed by multiplying by 3 the cost of a basic nutritionally adequate diet. This multiplier is based on a 1955 finding that poor people spend one-third of their income on food. Since then the poverty level has been readjusted annually to account for inflation. Using this official standard ($8,794 for one person under age 65, $13,738 for a family of three, and $17,603 for a family of four), 11.3 percent of Americans (31.1 million) were poor in 2000 (the statistics for this section are taken largely from U.S. Bureau of the Census, 2001).

While the government's arbitrary line dividing the poor from the near poor is the standard measure, it minimizes the actual extent of poverty in the United States. Critics of this measure argue that it does not keep up with inflation, that housing now takes up a much larger portion of the family budget than food, that there is a wide variation in the cost of living by locality, and that the poverty line ignores differences in medical care needs of individual families. Were a more realistic formula used, the number of poor would probably be at least 50 percent higher than the current official number (i.e., 17.0 percent or 46.6 million impoverished Americans).

The official number of poor people is also minimized because government census takers miss many poor people, an estimated 3 million in the 2000 census. Those most likely to be overlooked in a census live in high-density urban areas where several families may be crowded into one apartment or in rural areas where some homes are inaccessible. Some workers and their families who follow a harvest from place to place and therefore have no permanent home, as well as transients and homeless people, are also likely to be missed by the census. The inescapable conclusion is that the number of the poor in the United

Table 1 People and Families in Poverty by Selected Characteristics: 2000

Characteristic	Number Below Poverty	Percent
People		
Total	31,139,000	11.3%
Age		
Under 18 years	11,633,000	16.2
65 years and over	3,360,000	10.2
Nativity		
Native	26,442,000	10.7
Foreign born	4,697,000	15.7
Naturalized citizens	1,107,000	9.7
Not citizens	3,590,000	19.4
Race		
White, non-Hispanic	14,572,000	7.5
Black	7,901,000	22.1
Asian and Pacific Islander	1,226,000	10.8
Latino	7,155,000	21.2
Residence		
Inside metropolitan areas	24,296,000	10.8
Inside central cities	12,967,000	16.1
Outside central cities	11,329,000	7.8
Outside metropolitan areas	6,843,000	13.4
Type of Family		
Married couple	2,638,000	4.7
Female household, no husband present	3,099,000	24.7

SOURCE: U.S. Bureau of the Census (2000). "Poverty in the United States, 2000," *Current Population Reports* P60–214 (issued September 2001), p. 2.

States is underestimated because the poor tend to be invisible, even to the government. This undercount by the census and the out-dated formula used to determine the poverty line have policy consequences. The official numbers on poverty determine eligibility for federal programs such as Head Start, food stamps, and housing subsidies. When these numbers understate the real extent of poverty, they reduce the size of government programs to aid the impoverished (Maggi, 2000; Michael, 2001).

Despite these difficulties and the understating of actual poverty by the government's poverty line, the official government data provide useful information about the distribution of poverty in the United States. See Table 1 for a summary of the statistics used in this overview.

Racial Minorities

Income in the United States is maldistributed by race. In 1999, the median household income for Asian and Pacific Islander households was $51,205, compared with $44,366 for White households, $30,784 for American Indian and Alaska Native households, $30,735 for Latino households, and $27,910 for African American households. Not surprisingly then, in 2000 7.5 percent of

Whites were officially poor, compared with 10.8 percent of Asians and Pacific Islanders, 21.2 percent of Latinos, 22.1 percent of African Americans, and 25.9 percent of Native Americans and Alaska Natives. These summary statistics mask the differences within each racial/ethnic category. For example, Americans of Cuban descent, many of whom were middle-class professionals in Cuba, have relatively low poverty rates, whereas Puerto Ricans, Mexicans, and Central Americans have disproportionately high poverty rates. Similarly, Japanese Americans are much less likely to be poor than Asians from Cambodia, Laos, and Vietnam.

Nativity

In 2000, 4.7 million of the foreign-born individuals in the United States (15.7 percent) were poor. Of the poor foreign-born, 1.1 million were naturalized citizens and 3.6 million were noncitizens, with poverty rates of 9.7 percent and 19.4 percent, respectively. These official statistics do not include the 6-to-10 million undocumented workers and their families who enter the United States illegally to work at menial jobs.

Age

The nation's poverty rate for children under age 18 was 16.2 percent (11.6 million children) in 2000. For those children under age 6, the rate was 18.0 percent. Related children under age 6 living in families with a female head of household, no husband present, had a poverty rate of 50.3 percent, a rate more than five times that for their counterparts in married-couple families (9.0 percent). Although there are more White children in poverty, children of color are disproportionately poor. In 2000 the rate for White children under age 18 was 13.5 percent, compared to 33.1 percent of African American children and 30.3 percent of Latino children.

Contrary to popular belief, the elderly (those 65 and older) have a lower poverty rate (10.2 percent) than the general population (11.3 percent). The poverty rate increases with age—8.9 percent for those 65 to 74 years are poor, compared with 10.7 percent for those 75 and older. This is probably because women constitute an ever greater proportion of the elderly, and they have fewer Social Security and other retirement benefits than men do. For example, whereas only 6.6 percent of men 75 and older are poor, 13.4 percent of women in this age group are. Gender and race combine to make the economic situation especially difficult for elderly African Americans (a rate of 29.0 percent for women 75 and older) and Latino women (27.4 percent), compared with 12.1 percent of elderly White women.

Children have a higher poverty rate than the elderly because government policy provides programs such as Social Security and Medicare for the aged, and these programs are typically indexed for inflation. On the other hand, welfare programs targeted for poor children (e.g., Aid to Families with Dependent Children) have been eliminated or reduced, especially since 1980.

Gender

Women are more likely to be poor than men (13.2 percent compared with 10.3 percent) because of the prevailing sexism in society. The United States has a dual labor market, with women found disproportionately in lower paying jobs. In comparable jobs the female-to-male earnings ratio was 0.74 in 2000 (i.e., women earned 74 cents for every dollar earned by men). The relatively high frequency of divorce and the large number of never-married women with children—coupled with the cost of child care, housing, and medical care—have resulted in the high numbers of poor, single-female-headed families, with no husband present, being poor (24.7 percent of single-headed families in poverty, compared with 4.7 percent for two-parent families).

The **ascribed characteristics** (inherent characteristics over which individuals have no control, such as age, family of origin, race, and gender) of race and gender combine to increase the probability of poverty. An African American woman is almost two and a half times more likely to be poor as a White woman (26.6 percent compared with 10.9 percent); a Latino woman is almost twice as likely to be poor as a White woman (20.1 percent in contrast with 10.9 percent).

Place

Poverty is not randomly distributed geographically; it tends to cluster in certain places. Regionally, the area with the highest concentration of poverty in 2000 was the South (12.5 percent), compared with 11.9 percent in the West, 10.3 percent in the Northeast, and 9.5 percent in the Midwest. The South and West have the highest poverty rates because they have the largest concentrations of African Americans and Latinos and a relatively large number of recent immigrants. The five states with the highest average poverty rates from 1997 to 1999 were New Mexico (20.8), Louisiana (18.2), Mississippi (16.8), West Virginia (16.7), and Arkansas (16.4). Each of these states has a disproportionate number of racial minorities and a relatively high rural population.

In metropolitan areas, the poverty rate in 2000 was higher in the central cities (16.1 percent) than in suburban areas (7.8 percent). For those living outside of metropolitan areas, the poverty rate was 13.4 percent. One especially relevant trend regarding poverty in the central cities is that it is becoming more and more geographically concentrated (Massey, 1996). This spatial concentration of poverty means that the poor have poor neighbors, and the area has a low tax base to finance public schools and a shrinking number of businesses because businesses tend to move to areas where the local residents have more discretionary income. These factors mean a reduction in services and the elimination of local jobs. Moreover,

> Just as poverty is concentrated spatially, anything correlated with poverty is also concentrated. Therefore, as the density of poverty increases in cities . . . so will the density of joblessness, crime, family dissolution, drug abuse, alcoholism, disease, and violence. Not only will the poor have to grapple with the manifold problems due to their own lack of income; increasingly

they also will have to confront the social effects of living in an environment where most of their neighbors are also poor. (Massey, 1996:407)

Although poverty generally is more concentrated in cities, the highest concentration of U.S. poverty exists in four nonmetropolitan regional pockets where the economic engine has broken down: the Appalachian mountain region, where the poor are predominantly White; the old southern cotton belt from the Carolinas to the Louisiana delta, where the poor are mostly African American; the Rio Grande Valley/Texas Gulf Coast (Starr County, Texas, for example had a poverty rate of 43.8 percent in 1998), where the poor are largely Latino; and the Native American reservations of the Southwest.

Poverty is greatest among those who do not have established residences. People in this category are typically the homeless and migrant farm workers. The homeless, estimated at about 1 million, are those in extreme poverty. Migrant workers—adults and children who are seasonal farm laborers working for low wages and no benefits—are believed to constitute about 3 million. Latinos are overrepresented in this occupation.

The Working Poor

Having a job is not necessarily a path out of poverty. National data that include poor adults 16 years old and older—and excludes poor adults who are ill, disabled, retired, or in school—found that (1) almost two-thirds of all poor people work at some time during the year and (2) one-seventh of all poor people work full time for the entire year (summarized in Miller, 2000). Put another way, about 3 percent of all full-time workers are below the poverty line, with non-White, full-time workers one and one-half times more likely to be poor than White full-time workers (Barrington, 2000). Expanding this to the poor and the near poor, Congressman Bernard Sanders of Vermont has said: "30 percent of American workers earn poverty or near-poverty wages . . ." (Sanders, 2000:3). Despite working, these people remain poor because they hold menial, dead-end jobs that have no benefits and pay the minimum wage or below. About 10 million workers (70 percent of whom are adults) work for the minimum wage, yet this wage does not support a family. The federal minimum wage of $5.15 an hour adds up to only $10,712 for full-time work (it takes $8.20 an hour for a full-time worker to earn enough money to reach the poverty level for a family of four). Just above the minimum wage are more than 2 million Americans working in nursing homes; they earn, on average, between $7 and $8 an hour. The median wage of the estimated 2.3 million child-care workers is $6.60 an hour, usually without benefits. Workers in other occupations—such as janitors, health care aides, hospital orderlies, funeral attendants, and retail sales clerks—typically make similarly low wages (Reich, 2000).

Those working at poverty wages are disproportionately women and racial/ethnic minorities. One in three women earned poverty level wages in 1999, compared with one out of five men. Mishel, Bernstein, and Schmitt (2000) found that in 1999, one in three African American men, two out of five Black women and Latino men, and slightly more than half of Latino women worked at jobs that paid poverty wages. Compounding the economic woes of the poor is the fact that only three out of ten workers whose wages place them in the bottom fifth have employer-provided health insurance (Mishel, Bernstein, & Schmitt, 2000).

The Near-Poor

The **near-poor** are people with family incomes at or above their poverty threshold but below 125 percent of their threshold (i. e., with the official poverty line at $13,738 for a family of three, 125 percent of that number is $17,172. In 1999 some 12.0 million Americans were in this category, including 9.1 million Whites, 2.2 million African Americans, and 2.8 million Latinos.

The Severely Poor

Use of the official poverty line designates all people below it as poor whether they are a few dollars short of the threshold or far below it. In reality, most impoverished individuals and families have incomes considerably below the poverty threshold. In 1999, for example, the average dollar amount needed to raise a poor family out of poverty was $6,687 (i.e., the average family needed $6,687 in *additional* income to reach the poverty threshold). The per capita income deficit among people in families was $1,908. For poor individuals not in families, the average income deficit was $4,206 in 1999.

In 2000 some 39 percent of the poor population (12.2 million Americans) were **severely poor** (those people living *at or below half of the poverty line*). Some facts about these people who are among the poorest of the poor follow:

- Of these 12.2 million, 4.7 million (40 percent) were children under age 18.
- In this category, 3.36 million (28 percent) were African Americans, and 2.71 million (22 percent) were Latinos.
- Typically, the severely poor must use at least 50 percent of their meager income for housing.

The Dynamics of Poverty: The Likelihood of Moving Out of Poverty

Crucial to our understanding of poverty is knowing whether poverty is a transient phenomenon for individuals and families or whether it is a permanent condition. Research has shown that most poor people move above the poverty line in less than two years, whereas about 20 percent of those in poverty at any point in time have been poor for seven or more years (summarized by Rodgers, 2000). It is significant that most of those who escape poverty remain economically marginal and vulnerable. "Living on the margins means that if they encounter any economic adversity (e.g., sickness or injury, job loss, divorce), many of the former poor are in jeopardy of falling back into poverty" (Rodgers, 2000:66). Thus many people on the economic margins move in and out of poverty throughout much of their lives.

THE SOCIOLOGICAL IMAGINATION

This book is a collection of essays by people who have experienced poverty. Our goal in presenting them is to help the reader develop a better understanding of this social problem as described by those who are or who once were at the bottom of society's social and economic hierarchies. However, the fact that the essays are by poor people and their unique circumstances does not mean that we want the reader to understand poverty as a personal problem of people who have made choices that have led them to impoverishment. Rather we want the reader to view these essays with a "sociological imagination." C. Wright Mills (1916-1962) in his classic *The Sociological Imagination* (1959) wrote that the task of sociology is to realize that individual troubles are inextricably linked to social forces. Thus, by examining life circumstances, we can see how they are tied to the structure of society. The sociological imagination involves several related components:

- The sociological imagination is stimulated by a willingness to view the social world from the perspective of others.
- It involves moving away from thinking in terms of the individual and his or her problem, focusing rather on the social, economic, and historical circumstances that produce the problem. Put another way, the sociological imagination is the ability to see the societal patterns that influence individuals, families, groups, and organizations.
- Possessing a sociological imagination, one can shift from the examination of a single family to national budgets, from a poor person to national welfare policies, from an unemployed person to the societal shift from manufacturing to a service/knowledge economy, from a single mother with a sick child to the high cost of health care for the uninsured, and from a homeless family to the lack of affordable housing.
- Developing a sociological imagination requires a detachment from the taken-for-granted assumptions about social life and establishing a critical distance (Andersen & Taylor, 2000:10-11). In other words, one must be willing to question the structural arrangements that shape social behavior.

  > The sociological assumption that provides the basis for this critical stance is that the social world is human-made and therefore not sacred. Belief systems, the economic system, the law, the way power is distributed—all are created and sustained by people. They can, therefore, be changed by people. (Eitzen & Baca Zinn, 2001:8).

- When we have this imagination we begin to see the solutions to social problems not in terms of changing problem people, but in terms of changing the structure of society.

The Economic Downturn and Increasing Homelessness: A Brief Examination of How Social Forces Affect People's Lives

Several societal changes from 1999 to 2002 have converged to increase the likelihood of more people being pushed into poverty (in sociologese, macro forces impinge on the micro—individuals and families). To begin, there was an economic slump, which resulted in a rising unemployment rate. In September, 2001, seven million people were identified by the government as unemployed (the actual number is much higher), up from 5.5 million a year earlier. Then, on September 11, 2001, came the terrorist attacks on the World Trade Center and the Pentagon, which accelerated joblessness further (about 800,000 people were laid off in October and November). As consumer spending waned, companies downsized further. Low-skilled, low-income workers were disproportionately laid off, many with no savings and only a paycheck or two away from poverty. Moreover, the surge in layoffs created competition for the remaining jobs, leaving unskilled workers at a disadvantage as downsized skilled workers competed for ever fewer jobs. The plight of the marginalized and newly laid off was heightened by the ever smaller safety net provided by the federal and state governments (e.g., less money budgeted for Food Stamps and housing subsidies). Fewer than 40 percent of those who lost jobs received unemployment insurance because eligibility rules had grown tighter. Part-time workers, temporary workers, the self-employed, and those who had not worked long enough did not qualify. These tightened rules disproportionately affected lower-wage workers, especially women with young children (Reich, 2001). Moreover, for those who did qualify, the benefits varied widely by state (jobless benefits in Massachusetts were $477 a week compared with $230 a week in California) (Armour, 2001). On top of these forces that worked against the economically marginal, housing prices and rents, which had soared in the expansion of the 1990s, remained high even as the economy tumbled. The result was a dramatic decrease in affordable housing, with 5 million fewer apartments nationwide than were needed for people with the lowest incomes (National Low Income Housing Coalition, reported in Loven, 2001).

In the face of these societal forces, many vulnerable individuals and families were pushed into dire economic circumstances. One consequence was a sudden wave of homelessness (Belluck, 2001):

- A survey by the U.S. Conference of Mayors in late 2001 found that requests for emergency shelter in 27 cities had increased an average of 13 percent since 2000.
- According to the Coalition for the Homeless, in New York City the number of families in shelters increased by 50 percent from 1998 to 2001.
- A census conducted in Boston on December 10, 2001 found a record number of homeless in shelters and on the street.

- Shelter requests in Denver were 20 percent above those of the year before. The need for food among the homeless was 25 percent higher. An estimated 10 percent of the people needing food were not able to get it in adequate, nutritious amounts, and 30 percent of those seeking shelter were turned away for lack of space, leaving people to sleep in churches, cars, and bus stations (Hubbard, 2001).

In effect, then, a convergence of social forces created a housing crisis for more and more people. As examples of what this means at the personal level, consider the cases of John Swenson and Jim Kelling:

In Rhode Island, John Swenson, 44, took refuge at the Warwick shelter with his 10-year-old son, Michael, after he could not find work at his home in Hyannis, MA, and lost a part-time job cooking hamburgers in Warwick.

> "It's kind of late in life to be needing something," said Mr. Swenson, unemployed for the first time in 15 years and used to $14-an-hour jobs. "I knew there were shelters," he said, "and that's part of what kept me out of them. On the Cape, I helped paint the shelter in Hyannis. I went from painting a shelter to being in one." (Quoted in Belluck, 2001:3)

Jim Kelling, a Vietnam veteran, had worked for 30 years in various jobs dealing with computer programming (Wheeler, 2001). He was a database administrator at US West for two years, earning $25 an hour, before a merger with Quest resulted in his being laid off (Quest laid off another 7,000 workers in late 2001). He tried to survive on temporary data-entry jobs paying $10 an hour and even tried to earn a truck-driving certificate, but within a year he had lost his home and was forced to move into a homeless shelter.

John Swenson and Jim Kelling are homeless, not by choice but because macro forces worked against them and others in similar circumstances. The sociological imagination requires us to listen to people like them, understanding their situations in the light of the structural arrangements affecting their lives.

Let's embark on our journey by listening to poor people and learning from them about their lives and about the society that sustains such a high level of poverty. To do so requires that we develop and hone a "sociological imagination."

NOTES AND SUGGESTIONS FOR FURTHER READING

Andersen, Margaret L. and Howard F. Taylor (2000). *Sociology: Understanding a Diverse Society*. Belmont, CA: Wadsworth.

Armour, Stephanie (2001). "Tough Times for Laid-Off, Low-Income Workers." *USA Today* (October 30):1A–2A.

Barrington, Linda (2000). *Does A Rising Tide Lift All Boats*? New York: The Conference Board, Research Report 1271-00-RR.

Belluck, Pam (2001). "New Wave of the Homeless Floods Cities' Shelters." *New York Times* (December 18).

<www.nytimes.com/2001/12/18/national/18HOME.html>

Bergmann, Barbara R. (1996). *Saving Our Children from Poverty: What the United States Can Learn from France*. New York: Russell Sage.

Children's Defense Fund (2001). *The State of America's Children: 2001*. Washington, D.C.

Danziger, Sheldon, and Peter Gottschalk (1995). *America Unequal*. New York: Russell Sage.

Economic Policy Institute (2000). "The Myth of Economic Mobility." <http://www.epinet.org/books/swa2000/graph.htm>

Ehrenreich, Barbara (2001). *Nickel and Dimed: On (Not) Getting By in America*. New York: Metropolitan Books.

Eitzen, D. Stanley, and Maxine Baca Zinn (2001). *In Conflict and Order: Understanding Society*. 9th ed. Boston: Allyn and Bacon.

Frank, Robert H., and Philip J. Cook (1995). *The Winner-Take-All Society*. New York: Free Press.

Harrington, Michael (1962). *The Other America: Poverty in the United States*. New York: Macmillan.

Hubbard, Burt (2001). "Working Poor Losing Homes." *Rocky Mountain News* (December 19):5A.

Katz, Michael B. (1989). *The Undeserving Poor*. New York: Pantheon Books.

Kushnick, Louis, and James Jennings (eds.), (1999). *A New Introduction to Poverty: The Role of Race, Power, and Politics*. New York: New York University Press.

Loven, Jennifer (2001). "Entry-Level Pay Buys Little in Big Cities." Associated Press (December 24).

Maggi, Laura (2000). "The Poor Count." *The American Prospect* (February 14):14–15.

Massey, Douglas S. (1996). "The Age of Extremes: Concentrated Affluence and Poverty in the Twenty-First Century." *Demography* 33 (November):395–412.

Michael, Marie (2001). "What Is the Poverty Line?" *Dollars and Sense*, No. 233 (January/February):43.

Miller, Juanita E. (2000). "The Working Poor." Ohio State University Extension Fact Sheet <http://www.ag.ohio-state.edu/ohioline/hyg-fact/5000/ 5703.html>

Mills, C. Wright (1959). *The Sociological Imagination*. New York: Oxford University Press.

Miringoff, Marc, and Marque-Luisa Miringoff (1999). *The Social Health of the Nation: How America is Really Doing*. New York: Oxford University Press.

Mishel, Lawrence, Jared Bernstein, and John Schmitt (2000). *The State of Working America: 2000/2001*. Ithaca, NY: Cornell University Press.

Pollitt, Katha (2001). "Childcare Scare." *The Nation* (May 14):10.

Reich, Robert B. (2000). "The Great Divide." *The American Prospect* (May 8):56.

Reich, Robert B. (2001). "Lost Jobs, Ragged Safety Net." *New York Times* (November 12). <www.nytimes.com/2001/11/12/opinion/12REIC.html>

Rodgers, Harrell R., Jr. (2000). *American Poverty in a New Era of Reform*. Armonk, NY: M. E. Sharpe.

Sanders, Bernard (2000). "The 'Booming' Economy." *Sanders Scoop* (Spring):3.

Sen, Amartya (1995). *Inequality Reexamined*. Cambridge, MA: Harvard University Press.

Smeeding, Timothy M., and Peter Gottschalk (1998). "Gross-National Income Inequality: How Great Is It and What Can We Learn From It?" *Focus* (University of Wisconsin–Madison Institute for Research on Poverty), Volume 19 (Summer/Fall): 15–19.

Solow, Robert M. (2000). "Welfare: The Cheapest Country." *New York Review of Books* (March 23):20–24.

Schwarz, John E., and Thomas J. Volgy (1992). *The Forgotten Americans*. New York: W. W. Norton.

U.S. Bureau of the Census, *Population Reports* (2001) P60–214. Washington, D.C.: U.S. Government Printing Office.

Wheeler, Sheba R. (2001). "Homeless on Rise in Denver Since September 11 Attacks." *Denver Post* (December 20):3B.

Wilson, William Julius (1996). *When Work Disappears: The World of the New Urban Poor.* New York: Alfred A. Knopf.

Winfield, Nicole (2000). "U.S. Scores an 'F' on Efforts to Cut Poverty Among U.S. Women." Associated Press (June 8).

PART II

Theories of Poverty: Why Are the Poor Poor?

By design, we live in a society that is very competitive. After adding individual differences to handicaps/advantages of race, class, and sex, it turns out that some people run faster than others. Those who run very fast are highly rewarded. Those who run at an average pace do pretty well most of the time. For a variety of reasons, seldom if ever any fault of their own, some poor people and some homeless people . . . can't run fast at all.

ELLIOT LIEBOW (1993)

1

Individual/Cultural and Structural Explanations

How are we to explain the paradox of poverty amidst plenty? Why are there so many people poor in affluent America? There are two very different answers to these questions. One answer is that the poor are to blame because of their biological inferiority, their self-defeating culture, or their bad personal choices. The opposite answer places the blame on the structure of society. That is, some people are poor because society has failed to provide equal educational opportunities, because institutions have discriminated against women and minorities, because private industry has failed to provide enough good paying jobs, or because businesses have relocated to places where labor is cheaper. In this view, society works in a way that traps some people and their offspring in a condition of poverty.

INDIVIDUAL AND CULTURAL EXPLANATIONS

Innate Inferiority

Social Darwinism is the notion that social life is competitive, with the best and the brightest rising to the top of the social hierarchy and with the slow and the weak at the bottom. An early proponent of this idea, sociologist Herbert Spencer, said in 1882:

> [Poverty was nature's way of] excreting . . . unhealthy, imbecile, slow, vacillating, faithless members of society in order to make room for the 'fit,' who were duly entitled to the rewards of wealth. (Quoted in *The Progressive,* 1980:8)

The contemporary articulation of the "innate inferiority" argument is best illustrated by Richard Herrnstein and Charles Murray. They argue in their book *The Bell Curve* (1994) that the economic and social hierarchies found in society reflect a single dimension—cognitive ability, as measured by IQ tests. In effect society is a **meritocracy** (the stratification of people strictly by ability) developed by a sorting process. This reasoning assumes that people who are close in mental ability are likely to marry and reproduce, thereby ensuring social classes by level of intelligence.[1]

Cultural Inferiority

Most Americans intuitively accept the explanation for poverty known as "**the culture of poverty**." First articulated by anthropologist Oscar Lewis (1959), this theoretical position argues that many poor people, in adapting to their deprived condition, develop a way of life that keeps them poor. Individuals in this culture of poverty share feelings of marginality, helplessness, dependence, and inferiority. As a result, the poor have weak ego structures, a lack of impulse control, and a strong present-time orientation with relatively little ability to defer gratification and plan for the future. They are more permissive in raising their children, less verbal, more fatalistic, and less interested in formal education than the more affluent. Most significant, the different values and lifestyles that comprise the culture of poverty explain continued poverty because they are transmitted from generation to generation. They offer as proof the fact that the children of the poor are more likely than other children to be poor as adults. Moreover, these character traits are so ingrained that they will survive even if poor people suddenly leave poverty. Eventually, their values and behaviors would probably put them back into poverty because they would be impulsive and not oriented toward the future.

A variant of the culture of poverty explanation is the notion of **underclass** as presented by, among others, Auletta (1999), Lemann (1986), and Kaus (1992). The underclass is a relatively small group, perhaps 10 to 20 percent of the poverty population. The underclass are conceptualized as mired in poverty, living in the inner city, and, typically, non-White. This group adopts its own values and behaviors, which differ from those of mainstream society. The behavioral characteristics include a strong propensity for crime and other anti-social behaviors, welfare dependency, and out-of-wedlock births. A major facilitator of these deviant cultural values and behaviors is the welfare system, because it breaks down the work ethic by paying benefits as a matter of entitlement rather than as a consequence of work.

The most serious criticism of this notion that poverty results from the pathology of the weak is that it undercuts claims for collective obligation to design a better society (Robinson, 2001).[2]

---

[1]For critiques of the innate ability theories, see especially Ryan (1976); Gould (1994); Reed (1994); *Contemporary Sociology* (1995); and Fischer, et al. (1996).

[2]For critiques of the culture of poverty and underclass theories see especially, Leacock, E. (1971); Ryan (1970); and Reed (1992).

Most Americans believe that poverty is a combination of biological and cultural factors. Judith Chafel reviewed a number of studies on the beliefs of Americans and found that they

> view economic privation as a self-inflicted condition, emanating more from personal factors (e.g., effort, ability) than external-structural ones (e.g., an unfavorable market, racism). Poverty is seen as inevitable, necessary, and just. . . . (Chafel, 1997:434)

STRUCTURAL EXPLANATIONS FOR POVERTY

Michael Harrington, whose book *The Other America* was instrumental in sparking President Lyndon Johnson's war on poverty, said this:

> The real explanation of why the poor are where they are is that they made the mistake of being born to the wrong parents, in the wrong section of the country, in the wrong industry, or in the wrong racial or ethnic group. (Harrington, 1963:21)

In other words, the structural conditions of society are to blame for poverty, not the intellectual and cultural deficits of the poor.

Consider the way the U.S. economy works. The basic tenet of capitalism—that who gets what is determined by private profit rather than collective need—explains the persistence of some people's impoverishment. Because of the primacy of profits, employers continually keep wages and benefits as low as possible. The business community lobbies Congress, opposing the raising of the minimum wage, resists efforts to allow maternity leave for workers, and argues against "equal pay for equal work" provisions for women. Typically, employers have tried to keep unions out of the workplace because unions organize to increase wages and benefits for workers. Business owners also make investment decisions to reduce costs by using new technologies, such as robots, to replace workers. Similarly, owners may shut down plants and shift their operations to foreign countries where wages are significantly lower. In short, the capitalist system is an economic system that perpetuates inequality.

Most good jobs require a college degree, but the poor cannot afford to send their children to college. Scholarships go to the best-performing students, but children of the poor usually do not perform as well as the children of the privileged. This underperformance by poor children results from poor diets and exposure to lead in early childhood, lack of enriched preschool programs, low expectations for them by teachers and administrators, and systems of tracking by ability as measured on class-biased examinations. Problems in learning and test-taking also arise when English is their second language. Because poverty is often concentrated geographically and because schools are funded primarily by the wealth of their district, children of the poor usually attend inadequately financed schools. All these acts result in a self-fulfilling prophecy—the poor are not expected to do well in school, and many do not. They fall behind the children of privilege in developing skills for college and marketable skills that lead to success in jobs. Because these students are failures as measured by objective indicators (such as the disproportionately high number of dropouts and discipline problems and the very few who desire to go to college), the schools feel justified in their discrimination toward the children of the poor.

The traditional organization of work and education in U.S. society deprives racial minorities of equal opportunities for education, jobs, and income. As a result African Americans, for example, are half as likely to be wealthy and twice as likely to be poor as Whites. They are also twice as likely as Whites to be unemployed. Structuralists argue that these differences are *not* the result of the flaws in African Americans but rather result from historical and current discrimination in communities, schools, banks, and the work world. Similarly women typically work at less prestigious jobs than men do and, when they work at equal-status jobs,women receive less pay and have fewer chances for advancement. These differences are not the result of gender differences but are due to personal, social, and societal barriers to equality based on gender.

The structural explanation for poverty, then, rests on the assumption that the way society is organized perpetuates poverty, not the characteristics of poor people.[3] In the words of Claude Fischer and his colleagues:

> Inequality is not fated by nature, nor even by the "invisible hand" of the market; it is a social construction, a result of our historical acts. *Americans have created the extent and type of inequality we have, and Americans maintain it.* (Fischer et al., 1996:7)
>
> Theories of natural inequality cannot tell us why countries with such similar genetic stocks (and economic markets) as the United States, Canada, England, and Sweden can vary so much in the degree of economic inequality their citizens experience. The answer lies in deliberate policies. (Fischer et al., 1996:9)

Some examples of how the powerful and the privileged make policies that widen the economic gap between them and the poor include (Robinson, 2001):

- Over the last 20 years, the New Deal assumption that housing is a fundamental right has withered; budgets for subsidized housing have declined whereas gentrifying neighborhoods have been subsidized through government grants and redevelopment zones.
- The massive tax cut proposed by President George W. Bush and passed by Congress in 2001 gave 43 percent of the money to the wealthiest 1 percent of Americans while reducing social investments for health care, schools, and housing.
- Local governments provide subsidies to wealthy owners of professional sports teams to build luxurious stadiums. In Denver politicians put the corporate welfare stadium tax to a public vote, and lobbied for its pas-

[3]For critiques of the structuralist theories of poverty, see Murray (1984) and Mead (1992).

sage, but did not let the voters decide whether to extend that tax to help low-income renters and homeowners.

- Also in Denver, the city council approved a $65 million public subsidy for a luxury Hyatt hotel. At the same time these officials refused to honor requests to tie this subsidy to an agreement by the Hyatt to allow union organizing at the hotel.

> In such ways, the privileged make political choices that exacerbate the crisis of the vulnerable, justifying these choices with theories of the pathology of the poor. But politics can be transformed with the understanding that the problems of poverty are a human creation, amenable to human solution. (Robinson, 2001:G1)

The essays in this chapter cover a diverse range of people: a 40 year-old-white male; a 10-year-old Native American; a poor African American working family; residents of Appalachian coal country; a Black farmer; and a single, teenage mother in Denver. As you read them, think about how each essay contains the elements of one or more of the various explanations for poverty. Where does the blame for their condition lie—in themselves, in their culture, or in the structure of society?

NOTES AND SUGGESTIONS FOR FURTHER READING

Auletta, Ken (1999). *The Underclass,* updated and revised edition. New York: Random House.

Chafel, Judith A. (1997). "Societal Images of Poverty." *Youth and Society* 28 (June):432–463.

Contemporary Sociology (1995). "Symposium: The Bell Curve." Volume 24 (March):149-161.

Fischer, Claude S., Michael Hout, Martin Sanchez Jankowski, Samuel R. Lucas, Ann Swidler, and Kim Voss (1996). *Inequality by Design: Cracking the Bell Curve Myth.* Princeton, NJ: Princeton University Press.

Gould, Stephen J. (1994). "Curveball." *The New Yorker* (November 28):139–149.

Harrington, Michael (1963). *The Other America: Poverty in the United States.* Baltimore, Maryland: Penguin Books.

Herrnstein, Richard J., and Charles Murray (1994). *The Bell Curve Intelligence and Class Structure in American Life.* New York: The Free Press.

Kaus, Mickey (1992). *The End of Equality.* New York: HarperCollins.

Kushnick, Louis, and James Jennings (eds.), (1999). *A New Introduction to Poverty: The Role of Race, Power, and Politics.* New York: New York University Press.

Leacock, Eleanor B. (ed.), (1971). *The Culture of Poverty: A Critique.* New York: Simon & Schuster.

Lemann, Nicholas (1986). "The Origins of the Underclass." *The Atlantic,* Part I (June):31–55; Part II (July):54–68.

Lewis, Oscar (1959). *Five Families: Mexican Case Studies in the Culture of Poverty.* New York: Random House.

Mead, L. M. (1992). *The New Politics of Poverty: The Nonworking Poor in America.* New York: Basic Books.

Murray, Charles (1984). *Losing Ground.* New York: Basic Books.

National Law Center on Homelessness and Poverty (1999) <http://www.nlchp.org>

O'Connor, Alice (2001). *Poverty Knowledge: Social Science, Social Policy, and the Poor in the Twentieth-Century U.S. History*. Princeton, NJ: Princeton University Press.

The Progressive (1980). "Out of the Bottle." Volume 44 (August):8.

Reed, Adolph, Jr. (1992). "The Underclass as Myth and Symbol: The Poverty of Discourse about Poverty." *Radical America* 24 (January):21–40.

Reed, Adolph, Jr. (1994). "Looking Backward." *The Nation* (November 28): 654-662.

Robinson, Tony (2001). "Poor, Poorer, Poorest: The Politics of Poverty." *The Denver Post* (March 4):G1.

Ryan, William (1970). "Is Banfield Serious?" *Social Policy* 1 (November/December):74–76.

Ryan, William (1976). *Blaming the Victim*. Rev. ed. New York: Random House (Vintage).

Sidel, Ruth (1996). *Keeping Women and Children Last: America's War on the Poor.* New York: Penguin Books.

Poverty or At Home in a Car

Jackie Spinks

According to the National Law Center on Homelessness and Poverty (1999), the estimated number of homeless people on any given night is roughly 700,000, with up to two million people experiencing homelessness during a one-year period. While these numbers are staggering, what is it really like to be one of the severely poor, without a home? This essay was written by a 40-year-old White male at the "bottom of America's class system." Living in his car, eating at the local mission, he ruminates about being poor. As you read his essay, consider the following questions: Does the author reveal how he became poor? How has he adapted to his situation? Which theories of poverty (cultural/individual/structural) are reflected in this reading, and why? What role should society play, if any, in helping people like this author?

For seven months, my home has been my 1973 Dodge van. Life in a car is similar to life in a garbage can: trash filled, cramped, and with a smell to die from. Before that, I lived in a shed. I live here because I was kicked out of the shed.

SOURCE: Jackie Spinks, "Poverty or At Home in A Car," *Z Magazine* 9 (February 1996), pp. 57–60. Reprinted by permission of the publisher.

Guys, suited in black worsted, the authority clothes of caliphs-in-waiting, circle around me; avoiding eye contact.

Sometimes I wonder what's going through their minds when they see me? Do they ever wonder what's going through mine? Or do they give a damn? Or maybe they assume I don't have a mind? Or are they thinking, "Bum! Loser! What's the matter with you? There's work out there. It isn't a recession. You corks think you're too good to work. So okay, be a stiff, but why do I have to support you, and why do you dirty up nice streets? Where's your pride? Where's your work ethic?"

Yeah, I pretty much know them. But they don't know me from marmalade. So here's to making my acquaintance.

I'm one of those seedy guys at the bottom of America's class system, who knows that he's going to stay there. I'm not taken in by that "rags to riches" hooey our teachers, parents, media; in fact, the whole culture, fed us. I am 40-years-old, and have not earned more than $20,000 in my entire life, which means I've lived on less than $100 a month, for the past 20 years. So I know, after all this time, I'm not going anywhere.

Before I go on I'd better forewarn you: I'm not a curator of America's art of scrimping, a chronicler of poverty's abasements, or a bemoaner of its inequality. I'm just a down-and-outer trying to puzzle out my life, figure out how guys like me get where they are, especially when they don't have the excuse of alcohol or dope for their downward spiral.

I sold my Yamaha for $100 and bought this beater I'm living in for $75 dollars. Pretty lucky break. I got it for $75 because the interior was gutted, the windows broken, the roof rusted through. It was ready for the compactor, so nobody bid on it at the city auction. Thus, I have a home—my first in a long time.

I tinkered with it and got it moving. The reverse gear needs money to fix, so I can only go forward. The back brakes are bad (more money) and I worry driving, because of the brakes. So I don't drive it far. Besides, I have to hang onto my space, where my car's parked. But every once in a while I take a spin around the block just to keep my beater alive, as who knows when the local Praetorians will hustle me off. They'll give me a ticket for something, probably driving without registration. The constabulary knows and hates me. Not only am I an outcast, but being only five and a half feet tall, I'm an easy nab.

Mostly I stay where I am. What's important is keeping my corner. Three other guys have found my spot and are now parked near me. If I move my car, they'll grab it, as it is the best space, so I protect my claim. It's particularly valuable to me, as it's the only place where I can move forward and don't need to go into reverse.

It's a challenge, living in a crouched position in my van: minus toilet, water, telephone, electricity, and privacy. But the toughest part is the temperature: 95 degrees in the summer and freezing in winter.

I eat at the mission. Before I forced myself to check out the place (I was afraid of it), I lived on stuff like vanilla wafers, nachos, half-eaten pizzas, or sandwiches I found lying around.

Why the mission spooked me was, first, it was strange; and second, a lot of beefy guys with nothing to lose hung around the entrance. I had no weapon, no back-up, and there's always some guy hankering to beat up a little guy. On the one hand, poverty makes brutes, but on the other hand, people are poor because they're basically passive. Telling myself they all just want to be somebody and have you recognize it, I stiffened my mettle.

Nevertheless, just to avoid trouble, I smiled. I'm the kind of guy who smiles a lot, tells jokes, shows he's a regular guy. But I confess the adrenaline was pumping, and a rigid smile was shellacked on my face when I ambled cool, but not cocky, into the mission. It smelled like fat frying, a good smell, and the eating space was big—a high-ceilinged room with about 80 guys and 2 women sitting at long tables.

A lot of the guys were Mexican farmworkers, all talking in Spanish. They had serious faces. A couple years back, I climbed on a field bus and for about two weeks, picked strawberries. Those people were fast, about four times as fast as me. A little girl about 10 picked faster than I ever could—if I lived to be 100. I watched her to figure out her method. She'd kind of hold the bush and pull three or four berries at one reach. And did it with miniature fingers, that didn't crush the berries—such ingenuity wasted on a berry bush. She said she wanted a house and three kids when she grew up.

The food at the mission will never disturb McDonald's, but it fills me up and that's what counts. For instance, on my first day, they gave us two choices: bean soup or venison soup. We have a lot of deer hunters in this part of the country and most of them dislike venison, so the mission gets a lot of venison.

Most of the guys chose venison soup, but I chose bean soup. I'm a vegetarian, more by necessity than by conviction, but I think going without meat is a good idea. Besides, I have a soft spot for deer.

The mission, also, gets a lot of good pastries, the kind you buy at a bakery, usually about a week old. I eat a lot of doughnuts. You can take them with you when you leave. Sometimes we have ripe bananas. Sometimes we even have salad.

Often the food at the mission is spoiled, but nobody reports it, as it's better to be sick than risk closure. The bathroom, which is next to the kitchen, has been without toilet paper or paper towels for months. Four of the six toilets don't flush and nobody mops or uses cleanser, so you can figure it's a shopping mall of microbes, but nobody expects Trump Towers. Often I puke. Once, I did it for two days, but as one guy said, "How do you know it was the food at the mission. It could have been the flu."

Illness is for the rich. The poor that try to join that club, get two minutes and a swift heave-ho. It's been 20 years since I've been to a doctor. A few weeks back, I drove a guy with a severed finger to emergency, and sat with some mothers, who said they'd been waiting three hours. The mothers held their sick kids and scolded the others.

Many of the guys at the mission are here illegally. Nervous. Worried about trouble, too. Like me. And they're dirty. Like me. I can't speak Spanish, maybe a few words like "por favor, como esta, adios, and gracias." Wish I knew more. A big regret—flunking Spanish in high school. Their English is better than my Spanish, but still hard to understand. So mostly we just smile at one another.

My little forays into conversation with mission Saxons usually consist of mutual brag-a-thons. You'd think we'd all just dumped our portfolios with our brokers and decided to do a little slumming. And, if not that, we all have ships that will be tooting in, around the bend, any minute.

According to all of us displaced whites (I gas along with the best), we just need a little capital to get started. Some of the guys have pretty good ideas. One guy, a wood carver, thought he'd make Sasquash dolls and sell them at local fairs and bazaars. Another guy built a toilet for car dwellers, but fearing theft of his idea, was cagey in explaining how it operated. I listened up on that one. Then his paranoia took over, and he clammed up. Another had an idea for paper blankets.

It's all just talk. Nobody will ever do anything. To build up our self-esteem, we peg ourselves as 20th century pariahs, like frontier sheep farmers. We perceive ourselves as urban pioneers staking out claims, front runners of things to come. We carve out places, on the street, protect our turf, think of ourselves as prototypes of those guys who straggled over the Cascades without a penny to their names: sometimes desperate, often lonely, frequently surviving by their wits. They were giants. Stoics. Squatters like us. Our heroes. Deep in all of us, we mourn not having been born in the 18th century. Because we believe (probably incorrectly), that then poverty was a clean and honorable estate, not just wacko actions and fantasizing, like it is today.

How does a homeless, jobless, middle-aged man occupy his day? Well, first off, I sleep. Besides sleeping I post myself in the library. I arrive at opening time, ten in the morning, and stay most of the day. It's cool in there. Air conditioned. But temperature is only a minor reason I like the library. The main reason is the people. The people are quiet. No one bothers me or tells me to move on.

I read everything: religion, carpentry, mechanics, electrical, police training, political, biographies, nutrition, baseball novels, and a lot of gun, boat, news, and alternate press magazines. Sometimes I walk along the docks, examine the boats, imagine stowing away on a Russian cargo ship.

I scrutinize, with the eye of a tourist: buildings, factories, guys working—and I especially like to study bees. When I've exhausted looking, I sit in my car and read.

How do I live? Well, I use a can for my bowels and dump my loo near some pole beans and tomatoes in a garden nearby. I expected big beans and red, ripe tomatoes in that spot, but to my surprise they're limping along. One of my fellow car-dwelling aficionados said, "Living on the streets, you probably got toxic crap." I do my "miss congeniality" smile, but don't think it's funny. I don't think he meant it to be funny. I use recycled newspaper for toilet paper.

I mosey up to the college about once a month and sneak a shower in the college gym. I'm careful to use it when it's empty, shower as quickly as possible, just in case I get caught.

Probably the biggest problem for guys like me is water. I'm always looking for water. Every chance I get, I fill up a jug I carry everywhere in my backpack and haul it back to my Dodge. And that's strictly for drinking—not to wash hands, face, dishes, or clothes.

Apart from begging, the thing about poor people that riles rich people the most is our grime. Cleanliness is identified with goodness in America. It's next to Godliness. Dirt is evil. I think we're supposed to clean up and pretend we're real. I don't oppose cleanliness. It's that cleanliness for someone living in a car is a labyrinthine ordeal.

First, to wash and dry two loads of clothes—it costs about $5.00 total. Plus, you need a way to get to the laundry; plus, you need soap; plus, you need something to wear while you're washing the clothes.

But now and then a car-dwelling paisano does spiff up. Once cleaned up, he isn't too bad. I wonder why I've been going on ad nauseam about cleanliness and appearance? I must have some insecurity there. But then, so when aren't we insecure?

As for work, in my 20s when hope bloomed, I choo-chooed along, the little engine that could—but not anymore. Now, it's "tote that barge, lift that bail, wish to hell I could land in jail." I've worked for bosses that were drill sergeants with little concern about their oxen's safety who expected it to plow away at minimum pay. I got chemical burns on my hands that ate away chunks of the flesh, burns that still eat my fingers ten years later. I have no hope for anything better than a grinding $4.60 an hour.

A guy at the mission said, "I'd rather be poor than work. Why should I help make the rich richer?"

I'm anxious all the time. Until I found the mission, I worried constantly about where my next meal was coming from; I was anxious about my future; anxious about people. But mostly I was anxious that if—by some remote possibility I landed a decent job—that after so many years of non-work, with its absence of punctuality, drive, and concentration, I would fail again.

Hey, I sound like the mother of all belly-achers, bitching about contaminated food, cleanliness, jobs, pisspots, anxiety. I forget the other class has troubles, too; that they whip themselves, unmercifully, if some friend gains on them. That if their cash flow falls from ten million to two million, they worry they're finished and stumble home, get swacked, and feel like they'll never be a real player.

I heard some social worker-type talk about homelessness being a new deviant career. Sure am glad I have a career at last. Deviant or otherwise. Is homelessness better than being male, white, 40, and working at McDonalds? Yeah, I guess so, but then, Hell, McDonalds wouldn't hire a bum like me—even if he did have nice teeth.

The inner awareness I live with, and fight every day is that I'm nothing. To get rid of this feeling, I'd like to try Prozac. But two bucks for a pill, without a buzz. Then again, maybe I'm lucky. At some later date, they'd probably discover Prozac caused some kink, like edema of the brain, or elephantiasis, or more likely, an old stand-by like high blood pressure or cancer. They always find some pea to stick in our comfort zone. So I sleep Prozacless.

Because I sleep a lot, the books I've read label me depressed. I question that, as I've never contemplated suicide; but maybe I am depressed and don't know it. Or maybe I sleep because I'm bored. About the only truism about poverty in America in the 1990s is that it's one giant yawn.

Maybe that's what depression is—nothing to do. Rutsville. And doing the rut alone, at ebb tide, and feeling its all your fault. That everything's your fault. Feeling something's wrong with you because you would rather be poor then be under the foot of a master. So who wants to think. Better to sleep.

About our depression, whenever you channel onto a poor person's wave, male or female, what you discover is that—underneath the defenses, they're thinking, "There's something haywire with me. I'm worse than second-rate, I'm rotten. I can't do anything right. Nobody will ever hire a creep like me for any job that has a future. I'll never do anything that's even slightly smart."

To illustrate, a guy at the mission, came out from lunch and found the side of his Kawasaki dented. A few minutes before he'd been chair of the brag-a-thon. Now, you'd expect the guy to say, "I'm gonna kill the bastard who did this." But no, all this poor jerk could do was stare and mutter, "It's all my fault. I can't do anything right."

We all walked away, but it hit close. We understood.

In contrast to Americans, who place so much emphasis on individualism, people from other countries don't blame themselves. A Hindu cab driver, so black that he felt forced to reiterate three times, to our uninquiring minds, that he was Caucasian, told us about an experience he had here. He was sitting in a park and noticed several Americans going over to a marble protrusion, bending over it, and when water gushed out, drinking. Being thirsty, he went over to the protrusion, bent over it, but no water came out. He bent over it several times, still no water. Finally, he walked away, telling himself even the water over here in America is prejudiced.

Now that's what we considered an intelligent response, the kind we wish we could make.

It's like the world has a negative image of us, so we absorb that image, make it ours, and agree with the world, that, yes, because we aren't successes, we're slime.

Yeah, we're lazy. The dictionary definition of lazy is slow moving, resistant to exertion, slothful. And gadzooks, the perfect definition of depression.

Those active, employed people, who run around doing this and that, who talk briskly, walk with that high-stepping gait, who complain about have too much on their calendar, who, after a quick handshake, dash off to another appointment, are as strange to us as a bidet. We watch, wonder, and shuffle off, when we're around them. I guess they don't understand us either; the passive, the depressed, the underclass. Yet, we all share the same boat and it's gaining water. And maybe we know it better than they do.

Whatever. So, I sleep about 11 or 12 hours a day. Wish it were the sleep of the frazzled or anesthetized, but it's a half awake sleep. When awake I contemplate ways to make our world better. On good days, I dream of a revolution in concept.

On bad days I decide, "Blow 'er up. Start over."

As I lie here, watching the rain drip alongside the car window, I forage around for something positive to say about poverty. What I come up with is freedom. Yeah, I have freedom—no worry about stocks or kids or lovers. I click around the idea of freedom for a while and finally decide, next to a high IQ, freedom is the most over-rated quality any sociologist ever extolled.

The Armstrongs: An Oral History of a Homeless American Family

Steven VanderStaay

Steven VanderStaay introduces this selection in the following way: "The Armstrongs lived in Bellevue, a young, largely affluent city east of Seattle, until a medical emergency and the sudden loss of Mark's job forced the family to seek emergency housing. Since Bellevue has little emergency housing, the Armstrongs were advised to seek shelter in nearby Seattle. Eventually, the family was moved to a large public housing project in the city's Central District. Each morning they awake at 5:00 for the long bus ride back to Bellevue for work and school. Mark and Linda both work, as do their teenage children. Speaking to them, I am struck that they are the quintessential American family: hard-working, supportive, patriotic, loving. And now homeless. The Armstrongs' difficulties—underemployment, housing, grocery bills, health costs, insurance problems—mirror those of other homeless families driven from affluent communities. They are African Americans in their early thirties." As you read the following selection, what issues are raised by Mark's words about the minimum wage? The minimum wage has since been raised; is this enough to raise the Armstrong family out of poverty? How would individuals from the different theoretical perspectives analyze this family's situation?

MARK: I designed and built conveyor belts, and was good at it. I was making over $15 an hour. And I can go back there right now and get you a letter of recommendation from the company and let you read what they wrote about me. That in itself tells you what kind of worker I am.

The company went out of business. Bang! Didn't even know it was coming. I was between jobs three or four months. I could have found work right away if I wanted to make minimum wage, but I got pretty high standards for myself. I don't even want to make what I'm making now. We could barely afford rent then, how can we now? But when you got kids to feed and bills to pay, you have to do the best you can.

But minimum wage—that's insulting. I don't knock it for high school students. They're getting training, learning about working, making their pocket money. That's fine. But you take a person . . . I got six kids. $3.35, $4 an hour, I spend more than that wage in a day's time on a grocery bill. I mean you can accept some setbacks, but you can't tell a person, "I don't care if you've been making $15 something an hour, the minimum is what you've got to make now." If I hand you this letter, give you my resumé, my military record, show you the kind of worker I am, talk about my family, how can you degrade me by offering me the minimum wage?

Then we had trouble with the house we were renting. And, well, the biggest

SOURCE: Steven VanderStaay, *Street Lives: An Oral History of Homeless Americans* (Philadelphia, PA: New Society Publishers, 1992), pp.172–176. Reprinted by permission of the author.

part of it was hospital bills. My son had to have emergency surgery. Since the company was going out of business it let the insurance lapse, so I got stuck with the bill. Spent every penny we had saved and there's still fourteen hundred dollars on it. You would think by being medical that it wouldn't affect the credit, but it does.

Now I'm working with Safeway's warehouse. I work in the milk plant. Swing shift. Sometimes I'm off at 12:30, 1:00 at night, and then turn right around and go back at 8:30 the next morning. Yeah, it's hard sometimes. I'm not making half of what I used to. I'm a helper—I used to have people working for me. I'd worked my way up through the ranks. But like I was saying, you adjust, you do what you have to do. I'm the kind of person, I get with a company I want to stay, be a part of it. I like to get along with people and work, get my hands dirty. See something accomplished. I'm low man on the totem pole but I'll stay and work my way up.

The warehouse, it's refrigerated on one end and kind of hot on the other. They make their own milk cartons out of plastic so you have to deal with heat and cold. You have to know how to dress 'cause you're dealing with both extremes.

LINDA: I've been a custodian, nurse's aide; now I work at K-Mart. I still have to bus back to the East side [Bellevue] every day. It's okay but I'm looking for something else. You know, it's $4 an hour, and there's no benefits, no discounts at the store, nothing like that.

And I'm in school now, too. I'm going for business training, probably computers or administration. When school starts I'll either bring the little ones there with me or have one of the older ones bring them home.

Working full-time and going to school. Six kids, seventeen on down to twelve. Three in high school, three in grade school. Two of them work at Jack in the Box. They've been working the same shift but my oldest, he's on the football team, so he might be working at a different time than my daughter. And then there's the church, and those football games. Yes we're busy! Just an all-American family. One that's hit a string of bad luck, that's all.

The hardest thing is getting up early enough to bus back over there. As soon as school gets started that's really going to be a problem. It might be a couple of hours, both ways. And if they find out our kids are living here they'll want them in school in Seattle. But they like the schools there and I like them. They're better. And that's where we've lived, that's where we work.

But we get by. The kids, they cook, they clean, they wash and iron their own clothes. And the older ones, they all work. We're so proud of them. Oh, we have the same problems everybody else has, with teenagers and so forth. But we get through 'em. Just thank God they're not on drugs. That's the biggest problem here.

MARK: When we had to move and lost the house, when I lost my job, we told the kids the truth, the flat out truth. With no misconception; none whatsoever. Kids are not dumb. If you lie to kids, why should they be honest with you? They know exactly what we're going through and they know why.

Same thing when we moved here—six kids, three rooms, writing all over the walls, the drugs and crime. We tried to avoid the move but we didn't have any choice. They knew exactly where we were moving to, as best I could explain it. We told them we didn't want to come, but if it came down to it we were coming. And we did.

Now my worst fear . . . there's so much drugs in this area. And people think every apartment in the projects is a drug house. They knock on the doors, knock on the windows—they stop me out there and ask where it is. It's here, so close to us all the time. And all the shooting and fighting . . . You can look out the window any given night and see the police stopping people and searching everyone.

If I can't look out my door and see my kids, I send for 'em. And I'm afraid when I can't see 'em. 'Cause when they get to shootin' and fightin' and carryin'-on, a bullet don't got no names on it. Sometimes when I come in from work, three, four o'clock in the morning, I wonder just when they're going to get me. But my worst fear, my worst fear is the kids.

LINDA: Over in Bellevue they think if you can't afford it then you shouldn't be there. You know, who cares if you work there.

The first house that we had, we were the first blacks in the neighborhood. When I moved over there I said, "Where the black people?" [laughs, then moves her head from side to side as if searching] . . . no black people? Then the neighbors, they got to looking, came out, they were surprised, too. "Oooh, we got black people over here now" [laughs]. The kids were the only black kids around.

MARK: People don't want to rent to a family. And you know the kind of rent they're asking over there in Bellevue, that's not easy to come up with. And you need first, last month's rent, security. . . . And then people automatically assess, they stereotype you. Maybe sometimes it's 'cause we're black—I'm not saying this is true, I'm saying that sometimes I *felt* that the reason we didn't get a place was because we were black. But most of the time it's the family. People would rather you have pets than kids these days.

One guy, he had six bedrooms in this house. But he didn't want a family. Why would you have six bedrooms if you didn't want to rent to a family? May not be legal, but they do that all the time.

Now there is some validity in what they say about children tearing up things. But the child is only as bad as you let him be. You're the parent, he's going to do exactly what you let him do and get away with. If my kids tear something up, I'll pay for it. But me, I tell my kids that if I have to replace something they've destroyed, then one of their sisters or brothers isn't going to get something they need. And when they do something they answer to me.

I'm not bitter . . . I mean I'm somewhat so. I'm not angry bitter. It's just that I don't like dragging my kids from one place to the next, and I don't think we've been treated right. We had to take places sight-unseen, just to get 'em. We paid $950 a month, and during the wintertime $300, $400 a month for electric and gas bills. Then bought food, kept my kids in clothes. How you supposed to save to get ahead with all that?

And the house, when we moved in the landlords said they'd do this and that, fix this and that. Said we would have an option to buy it. We said "Okay, and we'll do these things." We had an agreement.

We never got that chance to buy, and they never fixed those things. But we kept paying that $950 a month. They had a barrel over us: we needed some place to go. And they made a small fortune those years. A month after we moved out we went by: all those things they wouldn't do were done.

Before that the guy decided to sell his house, just like that, and we had to move. It was December, wintertime. For a while we were staying with her mother in a two-bedroom. Nine people. We had to be somewhere so we took that second place before we had even seen it.

Everybody has to have a place to live. And people will do what they have to do to survive. A lot of things that you see going on around here are for survival [he sweeps his hand, indicating the housing projects]. I'm not taking up for them, there's a lot of things happening here that I oppose. But where there's a will there's a way, you know.

Worlds Apart

Cynthia M. Duncan

This selection was taken from a book about the persistence of rural poverty. Here the author takes a look at Blackwell county, an Appalachian coal county plagued by poverty and unemployment. In Blackwell county, only one-half of working-age men and one-quarter of working-age women are employed. In 1989, about 50 percent of the households in the county had incomes below $15,000. The selection is interspersed with the words of the residents there and their feelings about poverty and public assistance. As you read the selection, reflect on the terms "deserving" and "undeserving," "haves" and "have-nots." How do the words of the residents reflect the culture of poverty perspective? How would structural theorists respond to this selection?

Dependence on public assistance is widespread. Volatile mining employment combined with high levels of coal-related disability has meant that receiving public assistance is widely accepted and not stigmatized. When coal miners are laid off for several months they get food stamps and unemployment compensation, and their children are eligible for subsidized youth employment programs and other opportunities for the disadvantaged. In fact, many young women and men from families with middle-class incomes receive subsidized housing, training opportunities, food stamps, and even AFDC in some cases. They and their family and friends view their own welfare receipt as different from the dependency of the first-of-the-monthers, whom they assume have no interest in working. A teenage mother who is from a well-placed family but receives AFDC, food stamps, Medicaid, housing assistance, as well as subsidized child care and transportation while she attends college, explained, *We look down on people for whom receiving assistance is all they're going to do with their lives. My husband and I are not always going to have to have assistance because we're going to school. I think people realize that with us. My parents told me that there was nothing wrong with us getting assistance because we are paying for it in all actuality—because we pay our taxes into it and everything. We're not going to be on it all of our life. For me, personally, it would be hard not to look down on those who that is all they do and don't have any ambition.* Another welfare recipient whose parents have good jobs distanced herself from those who have long depended on welfare, referring to them as *scum, the bottom of the barrel, people you don't want to associate with.*

The prevalence of long-term dependency dominates descriptions of social life in the county in the 1990s. Community residents use phrases like "huge gulf," "cliff's edge," and "giant gap" to describe the distance between the haves and the have-nots. Those with jobs characterize the groups as "those who work" and "those who draw." It is assumed that those "who draw" do so by choice. *People that want to work are the same as people that do work,* says one of the employed, *because they're still trying to work. And then there's people who don't want to work at all, never have and never will. We call them first-of-the-monthers because they come out of the mountains the first of the month with about ten kids and don't wash. When I worked at the grocery store, you could smell them coming. But they just draw food stamps and stuff like that. They live like that, and I guess that's the way they want to live.*

These days coal miners at the big mines are considered among the haves. *The people who are just regular coal miners*

SOURCE: Cynthia M. Duncan, *Worlds Apart: Why Poverty Persists in Rural America* (New Haven, CT: Yale University Press, 1999), pp. 37–38. Reprinted by permission of the publisher.

can bring home maybe thirty or forty thousand, in that bracket. And then you go from that, and the transition is like a cliff's edge, and it drops off and you got people who are very, very poor. A newcomer can hear the same story about ambition three or four times in one week, each time told as if it happened to the speaker personally: *Why, just the other day I asked a young boy from up in the hollows what he wanted to do when he grew up and he answered, "I want to draw." I said, "Well, that's great. What kind of artist will you be?" And the boy replied, "No, draw a check."*

Few doubt that there are families here who have accepted dependency—who know nothing else and whose livelihoods depend on discovering what programs are available at the welfare office or the Community Action Agency, when commodities are handed out or coal is made available, and what churches help with light bills or groceries when the stamps run out. A minister I first interviewed in 1990, several years after he had arrived from Florida, found his parishioners and the county's citizens in general surprisingly resistant to reaching out to the "country poor." *I'm talking to them—I have for a year or two—about a soup kitchen. They give me this story about how every Tom, Dick, and Harry would come in there, and you would have all kinds of misfits. I'm sure that's true, but we could work that out. We could weed that out. I think that it would be good for our church to start that process. It would help them. It would help us. I think the word would get out and we would have a lot of people come.*

When we talked a second time, two years later, he said he still prodded his congregation to do more, but he also had developed a strategy to avoid subsidizing the same families making the rounds to his church door every month seeking handouts. *Some days it's like you need a store counter out there to wait on people as they come in for a handout. Power bills. There's one right here. Someone left it on my desk—$268 power bill. And it was $300 two months before. A few weeks ago I had one woman whom I'd never seen before come by and I gave her $80, to be exact. Three others came that same afternoon, came because they heard. You see, they have a network. There are people in here day in and day out wanting money for power bills, groceries, gas money. I've never seen it to this extent anywhere I've ever been. What we're going to do is phase out helping people, believe it or not*—he smiles at the irony in his words—*when they come here, unless it's an obvious real need. What we're going to do is reach out and find people ourselves that we know need help, through our members and through other agencies. And just go to them and help them.*

Experiences like this undermine community trust and reinforce the commonly held opinion that the very poor scheme to manipulate the system. Because this minister's parishioners are the well-to-do county-seat residents who have no contact with the poor, they will have difficulty finding people whom they can be sure "need" help. The social isolation that keeps the haves out of contact with the have-nots means that all long-term poor are stigmatized and lumped together as an undeserving group. Those with good jobs often use the pronoun "they," and speak with disdain about the dependent poor. A public employee said, *They know the system. They know how to get food stamps. They know all this stuff. They may not be able to read and write, but knowing the welfare system is like a job with them.* And the prejudice and distrust works both ways—people who have work often feel there is resentment of those who "try." *Poor kids call people who try snobs. If your parents have the drive and initiative to give you more, and you dress a little better and have a little more, then you're a snob regardless what your individual personality is.*

The poor know they are seen as a distinct, inferior group. As one teenager from a family with a "bad name" describes the community, *There are the "good rich people" and the "bad poor people," and they are segregated.* These distinctions are family based, and rich and poor alike agree that those whose "Daddy never did any good" will have a hard time getting work. A coal miner's son with a good public job obtained through political connections explained the way family names matter. *A lot of times you can hear somebody's last name and before you even meet them, you've already got the idea that they're either a good person or they're sorry as can be. Everybody knows everybody's family names. If you're a certain family, then you're this way. But if you're from that family, then you're that way. There are last names that you would just immediately associate with being trouble or lazy—they're immediately in a class.*

Employers in the private sector agree, whether they are in the mines, stores, or fast-food restaurants. Even the manager at the Department of Employment says coming from a family with a bad name is a disadvantage for young people. *Those that have a family with a horrible name, when they come in, we know them, and they're not worth two cents. They're sorry as can be—stealing, selling dope, bootlegging, picked up for driving drunk, in and out of bankruptcy court.*

Those from the families with the bad names give the same account about discrimination against them. They say family background matters for housing: *Everybody around here knows everybody, and they know what family you come from. Now my family, they've always been a bad family. There are places we can't even rent a house, because of our last name. And that's just the way it is. You can't change it. You have to live with it.*

I Dream A Lot

Ah-Bead-Soot

There is a high concentration of poverty on Native American reservations. For example, half of all children under the age of 6 living on reservations are below the poverty line. This essay is written by a ten-year-old girl on the Puyallup Indian Reservation, which is located in the far northwest corner of the United States. She discusses her dreams and goals and what it is like to attend a reservation school. As you read this selection, think about the way assimilation (the process by which individuals or groups voluntarily or involuntarily adopt the culture of another group, losing their original identity) has affected Ah-Bead-Soot. How could you use this essay to argue both culture of poverty and structural theories?

SOURCE: Ah-Bead Soot, "I Dream a Lot," *Messengers of the Wind: Native American Women Tell Their Life Stories*, Jane Katz (ed.). (New York: Ballantine Books, 1995), pp. 166–169.

You learn a lot about your culture. We have a circle every morning, and I'm one of the drummers, and we get to play and sing and stuff. Some kids are embarrassed to come up, but I think it's really fun. I'm proud to be Indian.

As an Indian, I have responsibilities. One time, I didn't want to get up for a parade, and I wished I were white so I wouldn't have to get up. But then I found out how much fun it is to take part.

My teacher is white, and I think she's leaning more about Indians than we're learning about whites. We're learning some of our old language, and our history. If you don't learn history, you'll make the same mistakes. Like if people keep on cutting the trees down the way they did in the past, there'll be no more trees. Whites do that. Indians do it too, for money.

The best thing about being in an Indian school is people don't tease you for being Indian. When we go into the white community, kids sometimes put their hands over their mouths and go "Wawawawaw."

I'm in a drama group which meets at the public school. Once we put on a play, and my dad came to a performance. He's Indian, he's a scientist. And this one white girl who I thought was my friend took one look at him and said to the other kids, giggling, "Look at that Indian with long hair!" Then, one of the girls I didn't like before because I thought she didn't like Indians, she stuck up for me saying, "Shut up! Just 'cause he has long hair, doesn't mean he's ugly." It was funny—I saw a different side of that girl.

My neighborhood is mostly Indian, but I have a couple of black friends. There was a little white girl who lived across the street from us—her name was Amber—and whenever I saw her, she'd pick a fight. Once, she started throwing rocks at me. Her mom came out and *she* started throwing rocks at me, then my sister came out to help me. We hit the mom in the head with a rock. I was going to apologize, but Amber pushed me down, and pulled my hair. My sister beat her up. Her parents never let her play with me again.

There are gangs here, and they're getting bigger and bigger. They sell drugs right near here on Portland Avenue. They're taking control. So many children are druggies—it's like peer pressure. When I was seven or eight, one of my brothers took an overdose of LSD and died, but the doctors brought him back to life. Another one of my brothers was hallucinating—he was walking on snakes. Now he's in jail for stealing car radios.

There was a crack house right next to the home of this tiny little girl who rides on my school bus. Things got so bad, she couldn't go outside. Finally, the police closed down the crack house. Now it's for sale, all the windows are fixed, the yard looks pretty. So people are trying to do something about the drugs.

You have to care about what happens. If you care only about yourself, it will just go on and on, and will affect future generations. Pretty soon everybody will be taking drugs. Drugs and alcohol, once you start you can't stop. There are all these relatives that my parents won't let me see because they use chemicals. My cousin is alcoholic. We don't have any alcohol in our house, but once he brought over three cans of beer—we found them in the backyard. Another time, his family, they got drunk and had a big fight and punched a hole in the wall of their house, and the police had to take them away.

There are homes I can't go to because they know I'll tell about the drinking. It's a secret. The parents drink because they're messed up, and they give alcohol to their children. The children who have parents who are drunks are the quietest. Sometimes they come to school with a black eye.

There are a lot of broken homes, with one parent gone, or both parents, and some of the kids have been taken from their families, and adopted. Sometimes kids drink because they miss somebody in their family who was part of their life. If chunks of your life are missing, you have to fill it with something.

My brother, Lah-huh-bate-soot, he's always in trouble with chemicals or the law, and that makes me sad. But then I just lie down and go to sleep. I dream a lot. I dream about how it would be if my cousin weren't here—he pushes me around. [Laughs.]

I dream about having everybody speak their tribal language. Your language is your way of life. Pretty soon even the grandparents will all talk English, and they will forget about the past, and then we won't hear the good stuff about when they were young.

We believe that when you die, you go to the other side of the world, and it's really pretty. It's all green, and the trees are green, and all the people who died are there in the spirit world. Grandpa is there, and I talk to him sometimes. He tells me how much he misses us.

I dream about being an actress some day, in theater and films. I know there aren't a lot of roles for Native Americans. But I think if you're really confident and have talent, you can get opportunities. Some people are snobs and meet one Indian and think all Indians are the same. But some people will take the time to get to know you and find out that you have ability.

I'm a fancy dancer. I go to powwows, I take part in dance competitions, and represent my tribe. Sometimes I win, and then I have to speak. Every year we have a big giveaway. You give everybody in the powwow something to show that you're proud of being what you are, and to thank those who helped you out. You give things you collect, like blankets. And you make things—beaded earrings, bracelets, special gifts for special people. My mom does beadwork—she taught me how.

Giving is important. If you don't give, you won't get anything in return. I'm not talking just about things, but about kindness. People have kindness in their hearts, and they share with others.

Family is important to me. My mom stays home a lot, she takes care of us, she feeds us good meals, and she makes us feel loved. I think how you raise your children is important. You have to take care of the child you brought into the world.

Feeling Trapped

Nikkie Thompson

The United States has the highest rate of teenage pregnancy of all developed countries. According to the Centers for Disease Control (2001), approximately 1 million teenagers become pregnant each year; 95 percent of these pregnancies are unintended. In addition to an increased risk of health problems to both the child and teenage mother, there is an increased risk for both to be poor. Going beyond the statistics, what is it like being a teen parent in poverty? Thinking sociologically, analyze this essay by a single, formerly teenage mother. What is her experience with the welfare system? What faults of the system does she point to? Which theories are best reflected in her essay and why?

SOURCE: Nikkie Thompson, "Feeling Trapped," People United For Families (PUFF) Poverty to Prosperity Newsletter (November 2000). Used by permission of the publisher.

Have you ever felt trapped? Not physically, but emotionally and financially. I have been a single parent for almost a decade and I have never been without stress. Being a parent is not easy, not to mention if you are a teenager. I could've done away with my responsibility, but I chose to take it on and become a woman.

Over the years I have been through homelessness, violence, and abandonment. Sometimes life should not offer so much strain. My first child clearly was not planned and my second was planned in the dream that daddy would always be there. Being on welfare with two kids has been no day at the beach.

My children are the most important thing to me. Can you believe that people ridiculed me and talked bad about me because I didn't have a job and lots of money? When I was growing up, my mom, grandma, and great grandma never had money. Thy managed some how to raise the people who raised me. They washed, scrubbed, wiped, bathed, and hard labored their asses off for pennies.

The rich people used the backs of the slaves to build the cities we live in now. I feel that times really have not changed. Let's start with the County Human Service Departments. Denver alone has around a $55,000,000 surplus. A family needing assistance for the TANF program (in some cases) has to complete a mandatory amount of work hours before they can receive any assistance. Or come to our job skill classes and receive no job skills at all.

Most of these programs teach you how to dress for an interview and how to get better self-esteem. And they encourage you into believing that you will be self-sufficient with a good job.

When you leave you will wonder what did I learn in these classes. And why do I have a low income, dead end ass job. Most people that leave these programs end up with a job in the $5.15–$8.50 range. That is not enough when considering housing, food, childcare, transportation, healthcare, clothing, phone, utilities, and taxes.

In my experience the government has a plan to keep the low-income and poverty level individuals just that. The system is set already to have us be dependent on it and not ourselves. They say "Give them welfare and make it hard to afford life." Then they'll say, "Take the welfare back and sell a false dream that they'll be self-sufficient from welfare to work programs."

We are these people just like my elders washing, picking, scrubbing, and cleaning up the city's filth. In today's society it takes money to make money. How can we as poor people get and keep anything with the GREED DOG on our back trying to take everything we get.

I am a 24-year-old working mom of two kids, making a little over minimum wage. I am stressed. I don't eat, sleep, or have time for my kids anymore. Just because I thought that I would be self-sufficient. I went to these programs made to "help." I ended up with a low paying job with hope of making more in the future. My money can never be saved because I don't make enough to save. Bills consume my whole check. Just from the little money I make, I am no longer eligible for any kind of assistance.

GREED is what this system is all about. How can we as the working poor ever afford the necessities of life? The minute we get some money it is gone. That sounds like TRAPPED to me!

No 40 Acres and a Mule: An Interview with a Displaced Black Farmer

Robert D. Bullard

The following selection is an interview with Gary Grant, an African American, North Carolina farmer. The selection points out the importance of land ownership to African Americans since the time of slavery. As you read this selection, think about the issue of institutional racism, which occurs when social arrangements and accepted ways of doing

SOURCE: Robert D. Bullard, "No 40 Acres and a Mule: An Interview with a Displaced Black Farmer," Environmental Justice Research Center (June 25, 1999). Reprinted by permission of the author.

things in society create a disadvantage for a racial group. How does Grant's story illustrate institutional racism? What components are evident from the structural perspective of poverty?

The federal government created the Freedmen's Bureau to provide assistance to former slaves. It also promised the former slaves parcels of land and the loan of a federal government mule to work the land. The federal government never lived up to its promise of "forty acres and a mule." Nevertheless, some African American farmers were able to buy or lease parcels of land under these programs. However, during Reconstruction under President Andrew Johnson, many of the powers and activities of the Freedmen's Bureau were dismantled and much of the land that had been leased to black farmers was taken away and returned to Confederate loyalists.

Despite open hostility, racial discrimination, and institutional racism practiced inside and outside of government, African American farmers were able to amass an impressive amount of farmland holdings. By 1910, they owned over 16 million acres of farmland. By 1920, there were 925,000 African American farmers. In 1999, African American farmers number dwindled to less than 17,000 and less than 3 million acres of land.

Racial discrimination practiced against African American farmers was never eradicated. In 1997, African American farmers brought a lawsuit against the USDA charging it with discrimination in denying them access to loans and subsidies. The lawsuit was filed in August, 1997 on behalf of 4,000 of the nation's 17,000 black farmers and former farmers. A Consent Decree was signed in January, 1999. The estimated cost of the settlement ranges from $400 million to more than $2 billion. To view the full court opinion click www.dcd.uscourts.gov/district-court.html

This interview was conducted with Gary Grant, a Tillery, North Carolina resident and a plaintiff in the Black farmers' lawsuit. Grant is president of the Black Farmers and Agriculturalists Association or BFAA. His family was forced out of farming in 1991. The interview was conducted by Robert D. Bullard, director of the Environmental Justice Resource Center, in June 1999.

QUESTION: What is your feeling about the black farmers settlement offered by a federal judge?

GRANT: It is a bittersweet victory. Sweet in the fact that we did succeed in the courts, having the USDA admit to discrimination and being certified as a class. Bitter in the fact that I still believe that this consent decree will ensure the demise of black farmers in two to five years. First of all, there is nothing in the document that returns land to us. In addition, there is nothing in the document that will pay off debt that has been incurred because of racist actions of USDA officers. Many of the farmers no longer owe USDA but owe private lenders. If their property is freed up, then the private lender will be able to come after it with less reserve.

QUESTION: What reservations do you have about how the settlement will be carried out?

GRANT: The settlement follows basically a two-track process. Track A is the track that will allow a farmer to go after $50,000 and have his debt written off and that's all. Track B is where a farmer provides records of bias from 1983–97, the period covered in the settlement. Black farmers have to prove with a "preponderance of the evidence" their case of discrimination. That will allow them to collect money from damages as well as to have debt written off to collect money for what they lost in not being able to farm. Track B would add up to more than $50,000. Getting into track B is just entirely too cumbersome for farmers to prove the discrimination in the atmosphere where racism ruled the day and people reacted based on their knowledge and understanding of how you act where racism is prevalent.

QUESTION: Which track do you feel the farmers will take?

GRANT: I really think that most of the farmers will be accepting the $50,000. It is virtually impossible to describe what people have been through and what most folk are saying. Most of them just want the USDA out of their lives. And when you hear the horror stories that people have to tell, you can understand why. Also, most of them have already been driven out of farming. They have lost their livelihood and a way of life. Once again, the government is asking the "victims" to prove discrimination. The burden of proof is on black farmers instead of the lawbreakers. That is not fair. That is not just. But, that's the American way. The government records are filled with examples where black farmers were systematically treated different from white farmers. Black farmers were routinely given less money for the same land than white farmers. They were also denied access to programs that aided white farmers. These were common practices. Fifty thousand dollars is not a lot of money. We are talking about a small business stolen, people's jobs, credit rating, and livelihoods ruined, life savings and investments taken, and spirits broken. No amount of money can repay the pain and suffering inflicted on black farmers. I wonder how much money the government would have offered us if we were white.

QUESTION: What guarantees do African American farmers have that the USDA will not allow the same thing to happen again?

GRANT: We have no guarantees. The government refused to put into the settlement document a clause that said it would enforce its civil rights policies. This means that they won't have a watchdog over them. Many of the farmers are displeased and frustrated. The farm advocate groups are most disturbed that for the long term there is nothing in the document that really helps us out. There is nothing in this document that guarantees that this kind of racism will not occur again within the USDA. To my knowledge, none of the USDA agents who perpetrated this injustice have been terminated. As a matter of fact, nobody was fired that I know of and some of them are getting promotions.

QUESTION: How has your family been hurt by the actions of the federal government?

GRANT: My community, Tillery, is a New Deal Resettlement community established in the 1940s. The federal government bought 18,000 acres of former plantation land and divided it up into forty to eighty-acre tracts and made it possible for black people to purchase that land. The black landowners have been the thorn in the side of the political power in Halifax County, North Carolina ever since, because we have not been dependent on them for our survival. Through our struggle, we've managed to save most of the land. Now the population of Halifax is about 52% African American and the community of Tillery is 99% African American. We still probably own 90% of the land, however white farmers are farming 98% of it.

Tillery had over 300 black farmers in the 1950s. Today, it has none. My family has been in foreclosure for 23 years and we continue to raise that issue. I am part of the class action because the USDA denied me the opportunity to assume my father's debt and to continue to operate our farm. My nieces and nephews have grown up in that 23 year time with a very bitter taste in their mouths and hearts about farming because they have seen the toll taken on my father and mother and my brother and his wife. We are not sure that any of them will enter farming. We are not even encouraging them that strongly, but it has brought us together much closer to the understanding of the power of the land.

QUESTION: How important is African American land ownership?

GRANT: Land ownership has to be a major theme that takes the African American community into the 21st century. As a people, we must understand the value and the power of land ownership. One generation removed from slavery, black folks were able to acquire more than 16 million acres of farmland. We have nearly lost it all. This land has been in my family for 52 years and another tract that was in jeopardy has been in my family for about 100 years. The only power that there really is in this country is land ownership, which produces economics, which is green stuff. Even for you to have money you have to own land first, so that the money has somewhere to be produced from. The constitution and the founding fathers believed that if you were not a member, if you were not a land owner you could not run for office, if you were not a land owner you could not vote. The lessons of the land have been there ever since the beginning. African Americans just seem to have problems with understanding and connecting to it. Also, as we lose this land no one is asking what happens to it and that is where we can also bring in the issue of environmental racism and environmental injustice. On much of this land is where the siting of polluting industries are being set.

QUESTION: How have you been treated by the federal courts?

GRANT: We were successful in keeping our case in court only because of the strong evidence pointing to racism practiced and condoned by USDA. It was time for America to come to grips with the ugly face of racism. America needs to know that a group of people was wronged. We were not satisfied with the January 1999 proposed settlement. On March 2, 1999, U.S. District Judge Paul Friedman held a fairness hearing to amend the lawsuit settlement. We brought close to 500 black farmers to Washington DC. We overflowed two federal court rooms. At that time, we had an opportunity to present to the judge our differences with the consent decree that had been filed. After that hearing, Judge Friedman made 14 recommendations based on what he heard in the courtroom that day. Our attorneys and the government attorneys only accepted four of those with a great modification and those four would not impact the actual actions of the consent decree.

QUESTION: What lessons can African Americans learn from the plight of black farmers?

GRANT: I think that the first lesson is the continuing fact that institutional racism is alive and well in this country. Black people still have to fight like hell to enjoy the rights that whites take for granted. I think the real lesson lies in what we can do as a people if we really will come together. Our case proved that we do have some political power. We got the statute of limitations set aside. We learned quickly that the media is largely controlled and our images manipulated in the headlines to suit the stereotypes of white people. Finally, it became clear that the African American community does not understand the real value and power of black landowners and black farmers. This is true for many of our black churches, civil rights organizations, political groups, colleges and universities, and professional associations. For the most part, we did not have a whole lot of mass support from the 40 million African Americans.

QUESTION: Why do you feel the black farmers' issues did not become a rallying point for many African Americans?

GRANT: Black farmers have produced more professional people than any other area of our society. We still are not getting widespread support from the black community. When you say black farmer and ask someone what image comes to mind, many will see a dirty, ignorant, barefoot, uneducated person. Many blacks see the stereotype. They don't understand that the black farmer has been a mathematician, a scientist, a meteorologist, a doctor, a veterinarian, and even a lawyer. Until we are able to destroy that stereotype, black farmers will always be misunderstood and unappreciated by our professional people.

QUESTION: What would you like to see black organizations, black institutions, and ordinary black citizens do to assist you in this struggle?

GRANT: First, I would like black institutions and black people to believe that the black farmers know what is needed. Second, I would also like them to contribute financially, morally, physically, and spiritually. Third, we need to begin a massive education program with our children on the importance of owning land. The historically black colleges and universities or HBCUs, and especially the Land Grant schools, need to get on board. The black farmers' struggle was a wake up call, and some of our institutions are still asleep. Our struggle challenged the plantation system. Many of our brothers and sisters do not want to stand up anymore and take a stand. Our land grant universities need to design outreach and research that encourage their students to work with black farmers and the black community.

QUESTION: What legacy would you like to see your struggle leave for generations?

GRANT: This country has not had to listen to black farmers because the black community has not said we are worth saving. I don't believe any of us will survive and progress unless we can come together around the central issue of the survival of black farmers. It is imperative that we maintain land ownership so we can make sure our food supply is not poisoned. Land ownership is economic power, political power, and is the only avenue that we really have to ensure our children a legacy.

REFLECTION QUESTIONS FOR PART II

The readings in this section contain elements of the different theoretical perspectives and are in no particular order. Analyze each article carefully and group them by individual/cultural and structural theories of poverty. Provide a rationale as to why you think one theory or another is best reflected in each reading.

1. How do the individuals "adapt" to their poverty?
2. Compare/contrast the readings in this section. Do you see any common themes?
3. Using your sociological imagination and the various readings as examples, explain how poverty is structured by race, class, gender, age, and location.
4. The theory you believe is more plausible will affect what you think the solutions are. What are the solutions to the different situations? How

do the readings compare in terms of solutions? What role should society play in helping the poor?

5. Select one particular essay, such as Cynthia Duncan's "Worlds Apart," and analyze it first *without* using a sociological imagination and then again *with* a sociological imagination.

PART III

Living on the Economic Margins

In the years since the media's discovery of race and poverty in America's inner cities, much has been written about . . . low-income housing projects. . . . [They] have become must-stops for anyone writing about the nation's so-called "underclass." Yet for all the ink and air time devoted to them, it is amazing how little we still know about the people who live there. . . . [R]arely do we get to know the people of the projects as anything other than sociological types, . . . one-dimensional portraits of third- or fourth-generation welfare mothers, violence-prone, drug-dealing gang members or street smart man-children living by their wits. Seldom do writers dare to look beyond the sociology and statistics . . . to see people as individuals rather than as examples of predrawn stereotypes.

SYLVESTER MONROE (1991)

While this book is devoted to understanding poverty as articulated by the poor themselves, we need to place their comments in the context of what is known about the consequences of living in poverty. This section examines these consequences regarding survival, discrimination, and difficulties in parenting.

2

Survival and Finances

As we have seen, 31.1 million people are below the poverty line, which is considered about $5,000 below the minimum subsistence budget for a hypothetical family of four. This means that the living conditions of the impoverished are, typically, substantially substandard. The poor must forgo medical and dental visits except for emergencies, put up with inadequate heat in the winter, and live in crowded apartments located in hazardous neighborhoods (i.e., neighborhoods with relatively high crime rates and a likelihood of environmental pollution levels significantly higher than those found in affluent neighborhoods).

The most significant fact is that the impoverished have inadequate diets. According to the U.S. Department of Agriculture, slightly fewer than 3 million families (2.8 percent of the nation's households) had at least one member who went hungry in 1999. Another 6 million households were on the edge of hunger ("food insecure") that year, meaning that they did not have assured access to adequate food at all times. Altogether, there were 27 million people, including 11 million children, who were hungry or at least food insecure in 1999 (Brasher, 2000). The elderly, especially older single women, are at risk for hunger. The National Policy and Resource Center on Nutrition and Aging estimates that 60 percent of those 65 and older are at high to moderate nutritional risk (reported in Weaver, 2000).

Inadequate resources often mean seeking the cheapest foods, not necessarily the most nutritious ones. As a result, the diet of the poor, typically, is too high in saturated fat (leading to heart problems), salt (a principal culprit in hypertension), and carbohydrates (leading to obesity and diabetes). Children in families below the poverty line are less likely than higher income families to have a diet rated as good (Federal Interagency Forum on Child and Family Statistics, 2000).

The minimum wage established by Congress is at least $3.00 an hour less than a living wage. Those who work at the bottom of the job-and-pay hierarchy rarely receive benefits such as health insurance.

As well as earning lower wages, having fewer work-related benefits, and paying a large proportion of their income for housing, the poor pay more than the nonpoor for many goods and services. The urban poor concentrated in the inner cities pay more for food and other commodities because supermarkets, discount stores, outlet malls, and warehouse clubs have bypassed inner-city neighborhoods. Because many inner-city residents do not have transportation to get to the stores where prices are lower, they must buy from nearby stores, giving those businesses monopoly powers.

Several forms of predatory lending take advantage of the poor (Hudson, 1996). Banks and Savings and Loans are rarely located in high-poverty-concentration areas. As a result, about 10 million poor families do not have bank accounts. This "unbanked" population is forced, then, to rely on a shadow banking industry composed of pawn shops, paycheck cashing businesses, and title brokers that loan money or cash checks to customers with poor credit at enormous costs. Title Brokers in Atlanta, for example, makes small loans to borrowers who use the titles of their cars for collateral at high interest rates. A loan of $600 must be repaid with $750 in one month or $1,172 in 90 days (Foust, 2000). Ralph Nader says that the effective interest rate of an auto title pawn is sometimes more than 900 percent (Nader, 2000). There are more than 9,000 payday lenders, who typically provide two-week loans for a 15 percent fee. Cash checking firms charge a fee equal to from 2 percent to 5 percent per check with a $2 minimum. According to *Business Week* there are "mortgage lenders who goad low-income homeowners into refinancing existing mortgages with new loans that carry such high rates, high fees, and hidden balloon payments that they virtually guarantee default—and foreclosure" (Foust, 2000:109). According to the National Consumer Law Center, as many as 600,000 Americans lose their homes each year because they were duped into bad loans (*USA Today,* 2000), and there are rent-to-own businesses that sell appliances and furniture to customers who cannot pay cash. Because of the outrageous interest rates, these consumers often pay from two to three times the cash price for their purchases. That the poor pay these high rates is not the result of their stupidity, but more often because of necessity. When you do not have the cash or the credit, sometimes you have no choice but to pay more for what you need.

The poor also pay relatively more in sales taxes for their purchases than the nonpoor. That is, the sales tax is a set tax, costing the same, theoretically, to every purchaser. Yet, charging the same rate means that the lower one's resources, the greater the proportion will be paid in tax, making it a **regressive tax** (an income tax where the rates increase with income is a **progressive tax**).

Thus, efforts to move federal programs to the states will cost the poor more, because state and local taxes tend to be regressive.

The facts, then, are clear: the poor have difficulty, sometimes incredible difficulty, in meeting their survival needs. The essays included here highlight these difficulties.

NOTES AND SUGGESTIONS FOR FURTHER READING

Brasher, Philip (2000). "Hunger in America Declines 24 percent Since '95." Associated Press (September 9).

Edin, Kathryn, and Laura Lein (1997). *Making Ends Meet: How Single Mothers Survive Welfare and Low-Wage Work.* New York: Russell Sage Foundation.

Federal Interagency Forum on Child and Family Statistics (2000). *America's Children: Key National Indicators of Well-Being, 2000.* Washington, DC: U.S. Government Printing Office.

Fine, Michelle, and Lois Weis (1998). *The Unknown City: Lives of Poor and Working-Class Young Adults.* Boston: Beacon Press.

Foust, Dean (2000). "Easy Money." *Business Week* (April 24):107–114.

Hudson, Michael (ed.). (1996). *Merchants of Misery: How Corporate America Profits From Poverty.* Monroe, ME: Common Courage Press.

Kotlowitz, Alex (1991). *There Are No Children Here: The Story of Two Boys Growing Up in the Other America.* New York: Doubleday Anchor Books.

Nader, Ralph (2000). "The Poor Still Pay More." *The Progressive Populist* (April 1):15.

Snow, David A., and Leon Anderson (1993). *Down on Their Luck: A Study of Homeless Street People.* Berkeley: University of California Press.

USA Today. (2000). "Foreclosers Rise as Lenders Take Advantage of Poor." (March 29):28A.

Weaver, Peter (2000). "Going Hungry: It Still Happens." *AARP Bulletin* 41 (June):9–11.

Ain't No Middle Class

Susan Sheehan

Originally published in The New Yorker, *this account is the true story of a working-class couple, struggling on the margins. The Mertens live in Des Moines, Iowa, with their two children. She is a nursing home aide, and he is a blue collar worker. Married for 22 years, their story reflects changing economic times, rising inflation, and growing debt. They are a prime example of a family with no safety net—one illness, injury, or job loss away from severe poverty. As you read this selection, think about their finances. What do you see for their future/old age? Is there a solution that will make their situation better? How would different theorists analyze their situation?*

At 10 o'clock on a Tuesday night in September, Bonita Merten gets home from her job as a nursing-home aide on the evening shift at the Luther Park Health Center, in Des Moines, Iowa. Home is a two-story, three-bedroom house in the predominantly working-class East Side section of the city. The house, drab on the outside, was built in 1905 for factory and railroad workers. It has aluminum siding painted an off-shade of green, with white and dark-brown trim. Usually, Bonita's sons—Christopher, who is 16 and David, who is 20 and still in high school (a slow learner, he was found to be suffering from autism when he was eight)—are awake when she comes home, but tonight they are asleep. Bonita's husband, Kenny, who has picked her up at the nursing home—"Driving makes Mama nervous," Kenny often says—loses no time in going to bed himself. Bonita is wearing her nursing-home uniform, which consists of a short-sleeved navy-blue polyester top with "Luther Park" inscribed in white, matching navy slacks, and white shoes. She takes off her work shoes, which she describes as "any kind I can pick up for 10 or 12 dollars," puts on a pair of black boots and a pair of gloves, and goes out to the garage to get a pitchfork.

In the spring, Bonita planted a garden. She and David, who loves plants and flowers, have been picking strawberries, raspberries, tomatoes, and zucchini since June. Bonita's mother, who lives in Washington, Iowa, a small town about a hundred miles from Des Moines, has always had a large garden—this summer, she gave the Mertens dozens of tomatoes from her 32 tomato plants—but her row of potato plants, which had been bountiful in the past, didn't yield a single potato. This is the first year that Bonita has put potato plants in her own garden. A frost has been predicted, and she is afraid her potatoes (if there are any) will die, so instead of plunking herself down in front of the television set, as she customarily does after work, she goes out to tend her small potato strip alongside the house.

The night is cool and moonless. The only light in the back yard, which is a block from the round-the-clock thrum of Interstate 235, is provided by a tall mercury-arc lamp next to the garage. Traffic is steady on the freeway, but Bonita is used to the noise of cars and trucks and doesn't hear a thing as she digs contentedly in the yellowy darkness. Bonita takes pleasure in the little things in life, and she excavates for potatoes with cheerful curiosity—"like I was digging for gold." Her pitchfork stabs and dents a large potato. Then, as she turns over the loosened dirt, she finds a second baking-size potato, and says "Uh-huh!" to herself, and comes up with three smaller ones before calling it quits for the night.

"Twenty-two years ago, when Kenny and me got married, I agreed to marry him for richer or poorer," Bonita, who is 49, says. "I don't have no regrets, but I didn't have no idea for how much poorer. Nineteen-ninety-five has been a hard year in a pretty hard life. We had our water shut off in July *and* in August, and we ain't never had it turned off even once before, so I look on those five potatoes as a sign of hope. Maybe our luck will change."

When Bonita told Kenny she was going out to dig up her potatoes, he remembers thinking, Let her have fun. If she got the ambition, great. I'm kinda out of hope and I'm tired.

Kenny Merten is almost always tired when he gets home, after 5 P.M., from his job at Bonnie's Barricades—a small company, started 10 years ago by a woman named Bonnie Ruggless, that puts up barriers, sandbags, and signs to protect construction crews at road sites. Some days, he drives a truck a hundred and fifty miles to rural counties across the state to set up roadblocks. Other days, he does a lot of heavy lifting. "The heaviest sandbags weigh between 35 and 40 pounds dry," he says. "Wet, they weigh 50 or 60 pounds, depending on how soaked they are. Sand holds a lot of water." Hauling the sandbags is not easy for Kenny, who contracted polio when he was 18 months old and wore a brace on his left leg until he was almost 20. He is now 51, walks with a pronounced limp, and twists his left ankle easily. "Bonnie's got a big heart and hires people who are down on their luck," he says.

Kenny went to work at Bonnie's Barricades two years ago, and after two raises he earns seven dollars and thirty cents an hour. "It's a small living—too small for me, on account of all the debts I got," he says. "I'd like to quit working when I'm 65, but Bonnie doesn't offer a retirement plan, so there's no way I can quit then, with 28 years left to pay on the house mortgage, plus a car loan and etceteras. So I'm looking around for something easier—maybe driving a forklift in a warehouse. Something with better raises and fringe benefits."

On a summer afternoon after work, Kenny sits down in a rose-colored La-Z-Boy recliner in the Mertens' living room/dining room, turns on the TV—a 19-inch Sylvania color set he bought secondhand nine years ago for a hundred dollars—and watches local and national news until six-thirty, occasionally dozing off. After the newscasts, he gets out of his work uniform—navy-blue pants and a short-sleeved orange shirt with the word "Ken" over one shirt pocket and "Bonnie's Barricades" over the other—and takes a bath. The house has one bathroom, with a tub but no shower. Last Christmas, Bonita's mother and her three younger brothers gave the Mertens a shower for their basement, but it has yet to be hooked up—by Kenny, who, with the help of a friend, can do the work for much less than a licensed plumber.

Kenny's philosophy is: Never do today what can be put off until tomorrow—unless he really wants to do it. Not that he is physically lazy. If the Mertens' lawn needs mowing, he'll mow it, and the lawn of their elderly next-door neighbor, Eunice, as well. Sometimes he gets up at 4:30 A.M.—an hour earlier than necessary—if Larry, his half uncle, needs a ride to work. Larry, who lives in a rented apartment two miles from the Mertens and drives an old clunker that breaks down regularly, has been married and divorced several times and has paid a lot of money for child support over the years. He is a security guard at a tire company and makes five dollars an hour. "If he doesn't get to work, he'll lose his job," Kenny says. In addition, Kenny helps his half brother Bob, who is also divorced and paying child support, with lifts to work and with loans.

Around 7:30 P.M., Kenny, who has changed into a clean T-shirt and a pair of old jeans, fixes dinner for himself and his two sons. Dinner is often macaroni and cheese, or spaghetti with store-bought sauce or stewed tomatoes from Bonita's mother's garden. He doesn't prepare salad or a separate vegetable ("Sauce or stewed tomatoes *is* the vegetable," he says); dessert, which tends to be an Iowa brand of ice cream, Anderson Erickson, is a rare luxury. Kenny takes the boys out for Subway sandwiches whenever he gets "a hankering" for one. Once a week—most likely on Friday, when he gets paid—he takes them out for dinner, usually to McDonald's. "It's easier than cooking," Kenny says.

Because Bonita works the evening shift, Kenny spends more time with his sons than most fathers do; because she doesn't drive, he spends more time behind the wheel. Christopher, a short, trim, cute boy with hazel eyes and brown hair, is one badge away from becoming an Eagle Scout, and Kenny drives him to many Scouting activities.

This summer, Kenny drove Eunice, who is 85, to the hospital to visit her 90-year-old husband, Tony, who had become seriously ill in August. After Tony's death, on September 12th, Kenny arranged for the funeral—choosing the casket and the flowers, buying a new shirt for Tony, and chauffeuring the boys to the private viewing at the funeral home. "Everyone was real appreciative," he says.

At around eight-thirty on evenings free from special transportation duties, Kenny unwinds by watching more television, playing solitaire, dozing again, and drinking his third Pepsi of the day. (He is a self-described "Pepsiholic.") Around nine-fifty, he drives two miles to the Luther Park nursing home for Bonita.

Bonita Merten leaves the house before 1 P.M., carrying a 16-ounce bottle of Pepsi (she, too, is a Pepsiholic), and catches the bus to work. She is dressed in her navy-blue uniform and white shoes. Since the uniforms cost 33 dollars, Bonita considers herself lucky to have been given a used one by a nurse's aide who quit, and she bought another, secondhand, for 10 dollars. Luther Park recently announced a mandatory change to forest-green uniforms, and Bonita does not look forward to having to shell out for new attire.

Bonita clocks in before one-forty-five, puts her Pepsi in the break-room refrigerator, and, with the other evening aides, makes rounds with the day aides. She and another aide are assigned to a wing with 20 long-term residents. "The residents have just been laid down on top of their beds before we get there," Bonita says. "First, I change water pitchers and give the residents ice—got to remember which ones don't want ice, just want plain water. We pass out snacks—shakes fortified with protein and vitamins, in strawberry, vanilla, or chocolate. They need the shakes, because they ordinarily don't want to eat their meals. While I'm doing that, the other aide has to pass out the gowns, washrags, and towels, and the Chux—great big absorbent pads—and Dri-Prides. They're adult snap pants with liners that fit inside them. We don't call them diapers because they're not actually diapers, and because residents got their pride to be considered."

At three-thirty, Bonita takes a 10-minute break and drinks some Pepsi. "We start getting the residents up and giving showers before our break and continue after," Bonita says. "Each resident gets two showers a week, and it works out so's I have to shower three patients a day."

One aide eats from four-thirty to five, the other from five to five-thirty. Until August 1st, Bonita bought a two-dollar meal ticket if she liked what was being offered in the employees' dining room. When the meal didn't appeal to her—she wouldn't spend the two dollars for, say, a turkey sandwich and a bowl of cream-of-mushroom soup ("I don't like it at all")—she either bought a bag of Cheetos from a vending machine or skipped eating altogether. On August 1st, the nursing home reduced meal tickets to a dollar. "Even a turkey sandwich is worth that much," she says.

The residents eat at five-thirty, in their dining room. "We pass trays and help feed people who can't feed themselves," Bonita says. "Sometimes we feed them like a baby or encourage them to do as much as they can." At six-thirty, Bonita charts their meals—"what percent they ate, how much they drank. They don't eat a whole lot, because they don't get a lot of exercise, either. We clear out the dining room and walk them or wheel them to their rooms. We lay them down, and we've got to wash them and position them. I always lay them on their side, because I like lying on my side. I put a pillow behind their back and a blanket between their legs. We take the false teeth out of those with false teeth, and put the dentures into a denture cup for those that will let us. A lot of them have mouthwash, and we're supposed to rinse their mouth. We're supposed to brush their teeth if they have them. After everyone is down, we chart. We check off that we positioned them and if we changed their liners. I'm supposed to get a 10-minute evening break, but I hardly ever take it. Charting, I'm off my feet, and there's just too much to do. Often we're short—I'll be alone on a hall for a few hours. The last thing we do is make rounds with the shift coming in. I clock out by nine-forty-five. Ninety-nine percent of the time, Kenny picks me up. When I had different hours and he'd be bowling, his half brother Bob picked me up, or I took a cab for five dollars. The bus is one dollar, but it stops running by seven o'clock."

Bonita has worked all three shifts at Luther Park. The evening shift currently pays 50 cents an hour more than the day shift and 50 cents less than the night shift, but days and nights involve more lifting. (In moving her patients, Bonita has injured her back more times than she can remember, and she now wears a wide black belt with straps which goes around her sacroiliac; she also uses a mechanical device to help carry heavy residents between their wheelchairs and their beds.) Bonita's 1994 earnings from Luther Park were only 869 dollars higher than her 1993 earnings, reflecting an hourly increase in wages from six dollars and fifty cents to six-sixty-five and some overtime hours and holidays, for which she is paid time and a half. This July 1st, she received the grandest raise that she has ever had in her life—75 cents an hour—but she believes there is a hold-down on overtime, so she doesn't expect to earn substantially more in 1995. Luther Park gives her a ham for Easter, a turkey for Thanksgiving, 10 dollars for her birthday, and 20 dollars for Christmas.

Bonita rarely complains about working at the nursing home. "I don't mind emptying bedpans or cleaning up the residents' messes," she says. She regards her job, with its time clocks, uniforms, tedious chores, low wages, penny-ante raises, and Dickensian holiday rewards, as "a means to a life."

Bonita and Kenny Merten and their two sons live in a statistical land above the lowly welfare poor but far beneath the exalted rich. In 1994, they earned $31,216 between them. Kenny made $17,239 working for Bonnie's Barricades; Bonita made $13,977 at Luther Park. With an additional $1,212 income from other sources, including some money that Kenny withdrew from the retirement plan of a previous employer, the Mertens' gross income was $32,428. Last year, as in most other years of their marriage, the Mertens spent more than they earned.

The Mertens' story is distinctive, but it is also representative of what has happened to the working poor of their generation. In 1974, Kenny Merten was making roughly the same hourly wage that he is today, and was able to buy a new Chevrolet Nova for less than 4,000 dollars; a similar vehicle today would cost 15,000 dollars—a sum that even Kenny, who is far more prone than Bonita to take on debt, might hesitate to finance. And though Kenny has brought on some of his own troubles by not always practicing thrift and by not always following principles of sound money management, his situation also reflects changing times.

In the 1960s, jobs for high school graduates were plentiful. Young men could easily get work from one day to the next which paid a living wage, and that's what Kenny did at the time. By the mid-80s, many of these jobs were gone. In Des Moines, the Rock Island Motor Transit Company (part of the Chicago, Rock Island & Pacific Railroad) went belly up. Borden moved out of the city, and so did a division of the Ford Motor Company. Utility companies also began downsizing, and many factory jobs were replaced by service-industry jobs, which paid less. Although there is a chronic shortage of nurse's aides at Luther Park, those who stay are not rewarded. After 15 years of almost continuous employment, Bonita is paid 7 dollars and 40 cents an hour—55 cents an hour more than new aides coming onto the job.

Working for one employer, as men like Kenny's father-in-law used to do, is a novelty now. Des Moines has become one of the largest insurance cities in the United States, but the Mertens don't qualify for white-collar positions. Civil-service jobs, formerly held by high-school graduates, have become harder to obtain because of competition from college graduates, who face diminishing job opportunities themselves. Bonita's 37-year-old brother, Eugene, studied mechanical engineering at the University of Iowa, but after graduation he wasn't offered a position in his field. He went to work for a box company and later took the United States Postal Service exam. He passed. When Bonita and Kenny took the exam, they scored too low to be hired by the Post Office.

Although 31 percent of America's four-person families earned less in 1994 than the Mertens did, Kenny and Bonita do not feel like members of the middle class, as they did years ago. "There ain't no middle class no more," Kenny says. "There's only rich and poor."

This is where the $32,428 that the Mertens grossed last year went. They paid $2,481 in federal income taxes. Their Iowa income-tax bill was $1,142, and $2,388 was withheld from their paychecks for Social Security and Medicare. These items reduced their disposable income to $26,417. In 1994, Bonita had $9.64 withheld from her biweekly paycheck for medical insurance, and $14.21 for dental insurance—a $620.10 annual cost. The insurance brought their disposable income down to $25,797.

The highest expenditures in the Mertens' budget were for food and household supplies, for which they spent approximately $110 a week at various stores and farmers' markets, for a yearly total of $5,720. They tried to economize by buying hamburger and chicken and by limiting their treats. (All four Mertens like potato chips.) Kenny spent about eight dollars per working day on breakfast (two doughnuts and a Pepsi), lunch (a double cheeseburger or a chicken sandwich) and sodas on the road—an additional $2,000 annually. His weekly dinner out at McDonald's with his sons cost between 11 and 12 dollars—600 dollars a year more. Bonita's meals and snacks at work added up to about 300 dollars. Kenny sometimes went out to breakfast on Saturday—alone or with the boys—and the meals he and his sons ate at McDonald's or Subway and the dinners that all four Mertens ate at restaurants like Bonanza and Denny's probably came to another 600 dollars annually. David and Christopher's school lunches cost a dollar-fifty a day; they received allowances of 10 dollars a week each, and that provided them with an extra 2 dollars and 50 cents to spend. The money the boys paid for food outside the house came to 500 dollars a year. The family spent a total of about $9,720 last year on dining in and out; on paper products and cleaning supplies; and on caring for their cats (they have two). This left them with $16,077.

The Mertens' next-highest expenditure in 1994 was $3,980 in property taxes and payments they made on a fixed-rate, 30-year, 32,000-dollar mortgage, on which they paid an interest rate of 8.75 percent. This left them with $12,097.

In April of 1994, Kenny's 1979 Oldsmobile, with 279,000 miles on it, was no longer worth repairing, so he bought a 1988 Grand Am from Bonita's brother Eugene for 3,000 dollars, on which he made four payments of 200 dollars a month. The Grand Am was damaged in an accident in September, whereupon he traded up to an 11,000-dollar 1991 Chevy Blazer, and his car-loan payments increased to $285 a month. Bonita has reproached Kenny for what she regards as a nonessential purchase. "A man's got his ego," he replies. "The Blazer is also safer—it has four-wheel drive." The insurance on Kenny's cars cost a total of $798, and he spent 500 dollars on replacement parts. Kenny figures that he spends about 20 dollars a week on gas, or about $1,040 for the year. After car expenses of $2,338 and after payments on the car loans of $1,655, the Mertens had $8,104 left to spend. A 10-day driving vacation in August of last year, highlighted by stops at the Indianapolis Motor Speedway, Mammoth Cave, in Kentucky, and the Hard Rock Cafe in Nashville, cost 1,500 dollars and left them with $6,604.

The Mertens' phone bill was approximately 25 dollars a month: the only long-distance calls Bonita made were to her mother and to her youngest brother, Todd, a 33-year-old aerospace engineer living in Seattle. She kept the calls short. "Most of our calls are incoming, and most of them are for Christopher," Bonita says. The Mertens' water-and-sewage bill was about 50 dollars a month; their gas-and-electric bill was about 150 dollars a month. "I have a hard time paying them bills now that the gas and electric companies have consolidated," Kenny says. "Before, if the gas was 75 dollars and the electric was 75 dollars, I could afford to pay one when I got paid. My take-home pay is too low to pay the two together." After paying approximately 2,700 dollars for utilities, including late charges, the Mertens had a disposable income of $3,904.

Much of that went toward making payments to a finance company on two of Kenny's loans. To help pay for the family's 1994 vacation, Kenny borrowed 1,100 dollars, incurring payments of about 75 dollars a month for two years and three months, at an interest rate of roughly 25 percent. Kenny was more reluctant to discuss the second loan, saying only that it consisted of previous loans he'd "consolidated" at a rate of about 25 percent, and that it cost him 175 dollars a month in payments. Also in 1994 he borrowed "a small sum" for "Christmas and odds and ends" from the credit union at Bonnie's Barricades; 25 dollars a week was deducted from his paycheck for that loan. Payments on the three loans—about 4,300 dollars last year—left the Merten family with a budget deficit even before their numerous other expenses were taken into account.

Except in a few small instances (according to their 1994 Iowa income-tax return, Bonita and Kenny paid H & R Block 102 dollars to prepare their 1993 return, and they gave 125 dollars to charity), it isn't possible to determine precisely what the rest of the Mertens' expenditures were in 1994. Several years ago, Kenny bounced a lot of checks, and he has not had a checking account since. Kenny exceeded the limits on both of their MasterCards a few years ago, and the cards were cancelled. Bonita has a J. C. Penney charge card but says, "I seldom dust it off." Now and then, Bonita went to a downtown outlet store, and if a dress caught her fancy she might put it on layaway. On special occasions, she bought inexpensive outfits for herself and for Kenny. Before last year's summer holiday, she spent seven dollars on a top and a pair of shorts, and during the trip Kenny bought a 75 dollar denim jacket for himself and about 50 dollars' worth of T-shirts for the whole family at the Hard Rock Cafe. One consequence of Kenny's having had polio as a child is that his left foot is a size 5½ and his right foot a size 7. If he wants a comfortable pair of shoes, he has to buy two pairs or order a pair consisting of a 5½ and a 7. Often he compromises, buying sneakers in size 6½. David wears T-shirts and jeans as long as they are black, the color worn by Garth Brooks, his favorite country singer. Christopher is partial to name brands, and Bonita couldn't say no to a pair of 89 dollar Nikes he coveted last year. The Mertens spent about 700 dollars last year on clothing, and tried to economize on dry cleaning. "I dry-clean our winter coats and one or two dresses, but I avoid buying anything with a 'Dry-clean only' label," Bonita says.

The Mertens' entertainment expenses usually come to a thousand dollars a year, but that amount was exceeded in 1994 when Kenny bought a mountain bike for every member of the family. The bikes (Bonita has yet to ride hers out of the driveway) cost 259 dollars apiece, and Kenny made the final payments on them earlier this year. This July, David rode Kenny's bike to the hardware store, and it was stolen while he was inside. Kenny yelled at David; Bonita told Kenny he was being too hard on him, and Kenny calmed down.

Bonita and Kenny don't buy books or magazines, and they don't subscribe to newspapers. (They routinely borrowed Eunice and Tony's Des Moines *Register* until Tony's death, when Eunice cancelled it.) They rarely go to the movies—"Too expensive," Kenny says—but regularly rent movies and video games, usually at Blockbuster. For amusement, they often go to malls, just to browse, but when they get a serious urge to buy they go to antique stores. Kenny believes in "collectibles." His most treasured possession is an assortment of Currier & Ives dishes and glasses.

The Mertens have never paid to send a fax, or to send a package via Federal Express, and they aren't on-line: they have no computer. They even avoid spending money on postage: Kenny pays his bills in person. Bonita used to send out a lot of Christmas cards, but, she says, "I didn't get a whole lot back, so I quit that, too." They spend little on gifts, except to members of Bonita's family.

Kenny knows how much Bonita loves red roses. Twenty-two years ago, he gave her one red rose after they had been married one month, two after they had been married two months, and continued until he reached 12 red roses on their first anniversary. He also gave her a dozen red roses when she had a miscarriage, in 1973, "to make her feel better." To celebrate the birth of David and of Christopher, he gave her a dozen red roses and one yellow one for each boy. And Kenny gives Bonita a glass rose every Christmas.

On a Sunday evening this summer, the four Mertens went to Dahl's, their supermarket of choice in Des Moines. They bought four rolls of toilet paper (69 cents); a toothbrush (99 cents); a box of Rice Krispies (on sale for $1.99); eight 16-ounce bottles of Pepsi ($1.67); a gallon of 1-percent milk ($2.07); a large package of the least expensive dishwasher detergent ($2.19), the Mertens having acquired their first dishwasher in 1993, for 125 dollars; two jars of Prego spaghetti sauce ($3); a box of Shake 'n Bake ($1.99); two rolls of film ($10.38), one for Kenny, who owns a Canon T50 he bought for 125 dollars at a pawnshop, and one for Christopher to take to Boy Scout camp in Colorado; a battery ($2.99) for Christopher's flashlight, also for camp; a pound of carrots (65 cents); a green pepper (79 cents); some Ziploc bags ($1.89); a Stain Stick ($1.89); a box of 2000 Flushes ($2.89); a package of shredded mozzarella ($1.39) to add to some pizza the Mertens already had in the freezer; and 12 cans of cat food ($3). Bonita bought one treat for herself—a box of toaster pastries with raspberry filling ($2.05). Christopher asked for a Reese's peanut-butter cup (25 cents), a bottle of Crystal Light (75 cents), and a package of Pounce cat treats ($1.05). All three purchases were O.K.'d.

David, who is enchanted by electrical fixtures, was content to spend his time in the store browsing in the light-bulb section. He was born with a cataract in his left eye, and the Mertens were instructed to put drops in that eye and a patch over his "good" right eye for a few years, so that the left eye wouldn't become lazy. Sometimes when they put the drops in, they told David to look up at a light. Today, David's main obsession, which apparently dates back to the eyedrops, is light. "We'd go someplace with David, and if there was a light with a bulb out he'd say, 'Light out,'" Bonita recalls. "We'd tell him, 'Don't worry about that,' and pretty soon he was saying, 'Light out, don't worry about that.'"

At 20, David looks 15. A lanky young man with coppercolored hair, hearing aids in both ears, and eye-glasses with thick lenses, he attends Ruby Van Meter, a special public high school for the city's mentally challenged. He reads at a fifth-grade level, and he doesn't read much. For years, the Mertens have been applying—without success—for Supplemental Security Income for David. In June of this year, when his application for S.S.I. was once again turned down, the Mertens hired a lawyer to appeal the decision. David has held a series of jobs set aside for slow learners (working, for instance, as a busboy in the Iowa statehouse cafeteria and in the laundry room of the local Marriott hotel), but he says that his "mood was off" when he was interviewed for several possible jobs this summer, and he drifted quietly through his school vacation. He will not be permitted to remain in school past the age of 21. If David could receive monthly S.S.I. checks and Medicaid, the Mertens would worry less about what will happen to him after they are gone. They have never regarded David as a burden, and although he has always been in special-education classes, they have treated him as much as possible the way they treat Christopher. Say "special ed" to Bonita, and she will say, "Both my boys are very special."

The Dahl's bill came to $44.75. When Kenny failed to take money out of his pocket at the cash register, Bonita, looking upset, pulled out her checkbook. She had expected Kenny to pay for the groceries, and she had hoped that the bill would be 40 dollars or less. But Kenny was short of money. "Aargh," Bonita said, softly.

Bonita didn't want to write checks for groceries, because she has other ideas about where her biweekly paychecks—about 400 dollars take-home—should go. Most of her first check of the month goes toward the mortgage—$331.68 when she pays it before the 17th of the month, $344.26 when she doesn't. Bonita likes to put aside the second check for the two most important events in her year—the family's summer vacation and Christmas. In theory, Kenny is supposed to pay most of the other family expenses and to stick to a budget—a theory to which he sometimes has difficulty subscribing. "I don't like to work off a budget," he says. "I think it restricts you. My way is to see who we have to pay this week and go from there. I rob Peter to pay Paul and try to pay Peter back." In practice, Kenny rarely pays Peter back. With his take-home pay averaging about 235 dollars a week, he can't.

When a consumer counsellor, who does not know the Mertens, was questioned about the family's current financial predicament—specifically, their 1994 income and expenditures—she made numerous recommendations. Among her suggestions for major savings was that the Mertens cut their food bills dramatically, to 5,400 dollars a year. She proposed stretching the Mertens' food dollars by drastically curtailing their eating out and by buying in bulk from the supermarket. She said that Kenny should get rid of his high-interest loans, and use the money he was spending on usurious interest to convert his mortgage from 30 years to 15. The way Kenny and Bonita were going, the counselor pointed out, they would not finish paying off their current mortgage until they were 79 and 77 years old, respectively. The Mertens' principal asset is 8,000 dollars in equity they have in their house. If the Mertens wanted to retire at 65, they would need more than what they could expect to receive from Social Security.

The counsellor had many minor suggestions for economizing at the grocery store. The Mertens should buy powdered milk and mix it with one-percent milk instead of buying two-percent milk. They should cut down even further on buying meat; beans and lentils, the counsellor observed, are a nutritious and less costly form of protein. She recommended buying raisins rather than potato chips, which she characterized as "high-caloric, high-fat, and high-cost."

The counsellor had one word for the amount—between 1,500 and 2,500 dollars—that the Mertens spent on vacations: "outlandish." Their vacations, she said, should cost a maximum of 500 dollars a year. She recommended renting a cabin with another family at a nearby state park or a lake. She urged the Mertens to visit local museums and free festivals, and go on picnics, including "no-ant picnics"—on a blanket in their living room.

Kenny and Bonita were resistant to most of the suggestions that were passed on to them from the counsellor, who is funded mainly by creditors to dispense advice to those with bill-paying problems. According to Kenny, buying a dozen doughnuts at the supermarket and then taking breakfast to work would be "boring." Bonita says she tried powdered milk in the mid-80s, when Kenny was unemployed, and the kids wouldn't drink it. She does buy raisins, but the boys don't really like them. Bonita and Kenny both laugh at the prospect of a no-ant picnic. "Sitting on the living-room carpet don't seem like a picnic to me," Bonita says.

Bonita surmises that the counselor hasn't experienced much of blue-collar life and therefore underestimates the necessity for vacations and other forms of having fun. "We couldn't afford vacations in the 80s, and if we don't take them now the kids will be grown," she says. Kenny reacted angrily to the idea of the boys' eating dried beans and other processed foods. "I lived on powdered milk, dried beans, surplus yellow cheese, and that kind of stuff for two years when I was a kid," he says. "I want better for my boys."

Kenny acknowledges that he tried to confine his responses to the consumer counsellor's minor suggestions, because he realizes that her major recommendations are sound. He also realizes that he isn't in a position to act on them. He dreams of being free of debt. He has tried a number of times to get a 15-year mortgage, and has been turned down each time. "We both work hard, we're not on welfare, and we just can't seem to do anything that will make a real difference in our lives," he says. "So I save 10 dollars a bowling season by not getting a locker at the alley to store my ball and shoes, and have to carry them back and forth. So I save 25 dollars by changing my own oil instead of going to Jiffy Lube. So what? Going out to dinner is as necessary to me as paying water bills."

Kenneth Deane Merten was born poor and illegitimate to Ruby Merten in her mother's home, outside Des Moines, on October 5, 1944; his maternal relatives declined to reveal his father's name, and he never met his father. Ruby Merten went on to marry a soldier and had another son, Robert. She divorced Bob's father, and later married Don Summers, a frequently unemployed laborer, with whom she had three more children. "Mr. Summers was so mean he made me stand up all night in the bed when I was eight years old," Kenny recalls. He has never hit his own sons, because "I know what it done to my life and I don't want it to get passed down." The family often moved in haste when the rent was due. Kenny attended eight or ten schools, some of them twice, before he completed sixth grade.

Kenny's mother died of cancer at 27, when he was 14. The three younger children stayed with Don Summers and a woman he married a month later. Kenny and Bob went to live with their maternal grandparents, and their lives became more stable. Even so, Kenny's school grades were low. "I had a hard time with math and science," he says. "Coulda been because of all the early moving around. I ain't stupid." He spent his high-school years at Des Moines Technical High School and graduated in 1964, when he was almost 20.

Two days later, he found a job as a shipping clerk for *Look* magazine. He kept the job until 1969, and left only when it became apparent that the magazine was cutting back its operations. He drove a cab from 1969 to 1972, drank too much, and did what he calls "some rowdy rambling." He had put much of that behind him when he got a job as a factory worker at EMCO Industries, a manufacturer of muffler parts and machinery bolts, in the fall of 1972, shortly before he met Bonita.

Bonita Anne Crooks was born on October 7, 1946, in Harper, Iowa. Her father, Cloyce Crooks, was employed all his working life by the Natural Gas Pipeline Company; his wife, Pauline, stayed home to take care of Bonita and her three younger brothers. Bonita was required to do chores, for which she was paid, and to deposit those earnings in a bank. She took tap-dancing lessons, wore braces on her teeth, and often went with her family on vacation to places like California and Texas. "Kenny's growing up was a lot worse than mine," she says. In 1965, Bonita graduated from a Catholic high school and became a nurse's aide, while living at home and continuing to bank her money. In 1971, she moved to Des Moines, and the following year she got a job as a keypunch operator for a large insurance company. Keypunching, however, proved too difficult for her (she couldn't combine accuracy with high speed), and she soon transferred within the company to a lower-paying position—that of a file clerk.

Bonita met Kenny in October 1972 on a blind date that had been arranged by a friend of hers. "I had been jilted by a younger man, and I knew Kenny was meant for me on our first date, when he told me he was born on October 5, 1944—exactly two years and two days earlier than me," Bonita says. She and Kenny fell in love quickly and were married in a traditional ceremony at a Catholic church in Harper on June 30, 1973. The newlyweds set off for Colorado on their honeymoon, but Kenny's car, a secondhand 1966 Pontiac Bonneville convertible, broke down, and the couple ended up in the Black Hills of South Dakota. When they were courting, Kenny had asked Bonita what sort of engagement ring she wanted. She had declined a "chunky" diamond, and said that matching wedding bands would suffice. "I suspected Kenny had debts," Bonita says. "I just didn't know how many he had until we got home."

The couple moved into a modest two-bedroom house. Bonita kept her file-clerk job after David's birth, in April 1975, but when she became pregnant with Christopher, who was born in November 1979, her doctor ordered her to bed. From the window of her bedroom, Bonita could see the Luther Park nursing home being built "kinda like next to my back yard." She didn't return to the insurance company, because her pay couldn't cover the cost of daytime care of two children. Kenny was working days at EMCO, so in June 1980 Bonita took a job on the 3-to-11 P.M. shift at Luther Park. She earned more there than she had as a file clerk. On some nights, Kenny drove a cab. He needed two jobs, because he regularly spent more than he and Bonita earned, just as he had overspent his own pay when he was single. Every year or two, he bought a new car. "I shouldn't have bought those new cars, but life with Don Summers made me feel completely insecure," he says. "Driving new cars gave me a sense of self-worth."

Kenny lost his job at EMCO at the end of 1983. He says that he had asked his supervisor for permission to take some discarded aluminum parts, and that permission was granted. But as he was driving off EMCO's premises with the parts in the bed of his pickup he was accused of stealing them. His supervisor then denied having given Kenny permission to take the parts. A demoralized Kenny didn't seek a new job for a year. He had already stopped driving the cab —after being robbed twice—and had started mowing lawns part time in the spring and summer, and doing cleanup work and shovelling snow in the fall and winter. Kenny's business failed—"There were too many unemployed guys like me out there." Many of his prized belongings were repossessed, among them a Curtis-Mathes stereo console. For two weeks in the summer of 1984, the Mertens were without gas or electricity or telephone service. They went on food stamps. Bonita felt guilty about going to work in air-conditioned surroundings while her husband and children were at home in the heat. Kenny felt humiliated when Bonita's parents visited their dark, sweltering house over the Fourth of July weekend. While Kenny has done better financially than most of his side of the family, it pains him that he hasn't done as well as Bonita's brothers, and that they regard him as a spendthrift and an inadequate provider. "When they get down on Kenny, I feel like I'm caught between a rock and a crevice," Bonita says.

Kenny's starting salary at EMCO had been seven dollars an hour. By the time he was terminated, it was eight-ninety-five an hour. In 1985, he found several jobs he liked, but none paid more than seven dollars an hour. One such job was with Bob Allen Sportswear, and he kept it until 1987, when he was let go during the off-season. He occasionally filed unemployment claims, and the family qualified for food stamps and received some groceries from food banks. During the rocky period between 1984 and 1988, Kenny tried to continue making payments on bills that he owed, in order to avoid having to declare bankruptcy, but his debts grew to the point where they exceeded his assets by "I think 12 or 13 thousand dollars"; his creditors—mostly finance companies—got fed up with him, and then he had no choice. The Mertens were able to keep their house and their '79 Olds. Going on food stamps didn't embarrass them—the boys had to eat, and they went off food stamps whenever Kenny had a new job—but the bankruptcy filing was published in the newspaper and made Bonita feel ashamed.

In 1989, after seeing an ad on television, Kenny enrolled in electronics courses at a local vocational school and borrowed 7,200 dollars to pay for his studies. His deficiency in math came back to haunt him, and he eventually dropped out. While at school, he had heard of an opening as a janitor at Ryko Manufacturing, an Iowa manufacturer of car washes. He eventually moved up to a factory job, working full time at Ryko in the early 90s for three years. Those years were happy ones. He got regular raises, and during the April-to-December busy season he earned a lot of overtime. In the summer of 1991, the Mertens flew to Seattle to visit Bonita's brother Todd. They had just enough money to cover one plane fare, and asked Bonita's brother Eugene to lend them the money for the three other tickets. Bonita took three months off that year; by then, she had worked full time at Luther Park for 11 straight years and needed a break. Kenny was proud to be the family's main provider, and wanted Bonita to stay home and take it easy.

In February 1993 Ryko fired Kenny Merten. His supervisors said that the work he did on the assembly line was neither fast enough nor of a sufficiently high quality. He was earning 11 dollars and 80 cents an hour—almost 30,000 dollars a year including overtime—when he was terminated. "In today's job market, first-rate companies like Ryko can afford to be selective," he says. "They want to hire young men."

Around the same time, Luther Park announced that it intended to expand. The nursing home offered the Mertens 39,000 dollars for the house they had lived in for 18 years. Kenny and Bonita accepted the offer, and were allowed to stay on, free of charge, for six months while they went house hunting. After they sold their house, it became apparent that they had been using it to supplement their income. The house they had bought for 14,800 dollars had appreciated handsomely in value, but they had kept re-mortgaging, and now they owed 29,000 dollars on it. As a result, they netted only 10,000 dollars from the sale. The purchase price of the Mertens' new home was 40,000 dollars. They spent 2,000 dollars from the sale of the old house on improvements to their new home, and this reduced the amount of the down payment they were able to afford to 8,000 dollars.

Kenny attempted to return to work at several of the companies where he had previously been employed, but they weren't hiring. It took him five months to find his current job with Bonnie's Barricades—far more arduous work, at lower wages than he had been paid at EMCO more than 20 years earlier. "I know I'll never be able to earn 11.80 an hour again," he says. "The most I can hope for is a seven-dollar-an-hour job that doesn't involve swinging sandbags. Maybe if I can come home less tired at the end of the day, I can handle an evening job."

This year did not get off to a good start for Kenny. In January, he hocked two rings that Bonita had given him for a hundred dollars, in order to pay a utility bill. Then, three months later, true to form, Kenny spotted two rings at a local pawnshop that he wanted Bonita to have—a 199-dollar opal ring and a 399-dollar diamond-cluster ring. He asked the pawnshop owner to take the two rings out of the showcase and agreed to make periodic 20 dollar payments on them until they were paid off.

Kenny was not worried about how he would pay for the rings, or how he would pay for the family's annual summer vacation. In September of last year, a few days after the Mertens returned from that summer's driving trip, his Grand Am was rear-ended. After the collision, in which Kenny hurt his back, he hired a lawyer on a contingency basis. The young man who had caused the accident had adequate insurance, and Kenny expected to be reimbursed for medical bills and lost wages. (He hadn't been permitted to lift heavy objects for several weeks.) He also expected the insurance company to pay a sizable sum—10 or 15 thousand dollars—for pain and suffering. Kenny's lawyer told him that he could expect the insurance company to settle with him by March. When the insurance money failed to arrive that month, Kenny's lawyer told him to expect an offer in April, then in May, and then in June. In early July, the lawyer said that he could get Kenny 6,500 dollars by the end of the month—just in time to save the Mertens' summer vacation. The insurance payment and the annual vacation had been the focus of Bonita's attention for seven months. "If you don't go on vacation, a year has gone by with nothing to show for it," she says.

Bonita wanted the family to travel to Seattle to visit Todd because he had a new home and she was eager to see it. The Mertens made meticulous plans for a driving trip to the state of Washington. They decided they would get up at 4 A.M. on Saturday, August 5th, and drive to Rapid City, South Dakota. They would visit Mt. Rushmore, and Kenny, who has an eye for landscapes, would take photographs of the Devils Tower, in Wyoming, at sunrise and sunset. They would arrive at Todd's home on Wednesday, August 9th, spend a few days there, and return to Des Moines, by way of the Mall of America, in Bloomington, Minnesota, on August 19th. Both Bonita and Kenny had arranged with their employers to take one week off with pay and one without.

Six days before their departure, however, their lawyer called with crushing news; the insurance payment would not be arriving until September. The following evening, Bonita injured her shoulder lifting a patient at the nursing home, but she was still determined to have her vacation. Although Kenny was behind on almost all his bills—he had just borrowed 75 dollars from David to pay a water bill—he went to a bank and to his credit union on August 2nd to borrow 2,500 dollars to cover the cost of the vacation, figuring he would pay off this newest loan from the insurance money in September. On the evening of August 2nd, Bonita reinjured her shoulder while helping another aide transfer a resident from her wheelchair to her bed.

Both the bank and the credit union turned Kenny down. Not only did he have too much outstanding debt of his own but he had also cosigned a loan on his half brother Bob's car. Without being able to borrow, the Mertens could not go on vacation. To make matters worse, Luther Park had sent Bonita to a doctor, and he informed her that she would require physical therapy three times a week for the next two weeks. The vacation would have to be cancelled. "When Kenny told me he'd been turned down for the loan, his jaw dropped about two inches," Bonita recalls. "Kenny was so shocked and disappointed for me that I couldn't be disappointed for myself."

The Mertens have had their share of disappointments, but they don't stay down long. On the morning they had set aside to pack for their trip, Bonita baked banana bread. That evening, after she finished work, Kenny took the whole family out to dinner. From there they drove to Blockbuster and bought two videos—"Sister Act 2" (David had loved the original) and a Beatles movie. They also rented two movies, and a video game that Christopher wanted. The boys spent the following week at their grandmother's. During the second vacation week, Bonita took David to the Iowa State Fair, in town. "Me and David really had fun together," she says.

Both Mertens spent a little money during the two weeks that they didn't go out West. Bonita made a payment to Fingerhut on a shelf that she had bought for David's room and on a game that she had bought him, and she finished paying Home Interiors for some mirrors, sconces, and a gold shelf that she had bought for her bedroom. "When I buy this stuff, I can see Kenny getting a little perturbed, but he doesn't say anything," she says. Later in August, the front brakes on Kenny's Blazer failed, and replacement parts cost about a hundred dollars. The labor would have cost him twice that much, but Eunice, the next-door neighbor, gave him some furniture that she no longer needed, and he bartered the furniture with a friend who is an auto mechanic. Kenny and Bonita agreed that driving with faulty brakes through the mountains on their way West would have been dangerous, so it was a blessing in disguise that they had been forced to remain at home.

On Friday, September 22nd, Kenny, feeling unusually fatigued, decided to take the day off from work. After lunch, he drove Bonita to their lawyer's office. The insurance company had agreed to pay Kenny 7,200 dollars. The lawyer would get a third—2,400 dollars—and Kenny owed 1,200 dollars in medical bills, so he would net 3,600 dollars. He had wanted more—to pay off more of his debts and bills—but this was three days after Bonita's lucky potato strike, and she was feeling optimistic. She persuaded Kenny to put the agony of waiting behind them and to accept the offer.

The next day, Kenny drove Bonita, David, and Christopher to the pawnshop. The proprietor, Doug Schlegel, was expecting them. At the cash register, Doug handed Kenny a small manila envelope with the opal ring inside. "Hey, Kiddo!" Kenny called out to Bonita as he removed the ring from the envelope. "Come here!"

Bonita tried to kiss Kenny, but he quickly moved away. "I love you," she said. After Bonita finished working the opal ring down the third finger of her left hand, checking to see whether it fitted properly, Doug told her, "You don't want to let it sit in the sun or put it in hot water."

"I know," Bonita said. "Opals are soft and touchy. They're my birthstone. I have one I bought for myself, but this is lots prettier."

Once the Mertens were back in the Blazer, Bonita asked Kenny, "Is the opal my birthday present?" Her 49th birthday was coming up in two weeks.

"It's a prebirthday present," Kenny replied. He didn't mention his plan to give her the more expensive ring—the one with the diamond cluster—for Christmas, provided he could make the payments in time.

"Thank you, Kenny. I love you," Bonita said.

"Sure," Kenny said. "You love to pick on me and drive me crazy."

Bonita touched Kenny's hand. "Leave me alone, I'm driving," he told her.

When Kenny stopped at a red light, Bonita said, "You're not driving now." But the light suddenly turned green.

Throughout the fall, Kenny Merten refused to fret over the very real possibility that he would have to file for bankruptcy again if he didn't get his financial house in order. He was thinking only as far ahead as Christmas—imagining himself putting the box that held the diamond-cluster ring for Bonita under the tree in their living room and marking it "Open this one last." Kenny predicts that when his brothers-in-law see the ring they will surely disapprove, but he doesn't care. "The rings shouldn't be in the budget, but they are," he says.

Kenny's mother's short life left him with a determination to marry once and to make that marriage succeed—something that few of his relatives have done. Bonita has often said that one reason she loves Kenny is that he surprises her every once in a while.

"Diamonds are a girl's best friend, next to her husband," Kenny says. "And Bonita's worth that ring, every bit of it. After all, she puts up with me."

Unafraid of the Dark

Rosemary L. Bray

Rosemary Bray is an African American woman who grew up on welfare. She is now a writer and former editor of The New York Times Book Review. *The following is an excerpt from her book, where she reflects on the survival techniques of her mother and other women in the neighborhood where she grew up. Bray's family and other families she knew lied to their welfare caseworkers and found ways to supplement their income without reporting it. What is it about the system*

SOURCE: Rosemary L. Bray, *Unafraid of the Dark: A Memoir* (New York: Random House, 1998), pp. 56–59.

that encourages this behavior? Is this taking advantage of the system? Why/why not? What role do the men play in this system? How would cultural and structural theorists explain the behavior of these families?

Everyone on our street knew who was on welfare and who wasn't, who had a man in the house, who didn't. Everybody also tacitly agreed it wasn't anybody's business, least of all the state of Illinois. It never occurred to anyone to reveal the circumstances of another family; the household in jeopardy could just as well be your own. This collective silence was a practical one; no one could possibly raise children on the money provided by AFDC. There were nearly 25,000 AFDC cases on the books in Cook County in 1960, accounting for about $4.3 million distributed to 105,000 women and children. That worked out to an average of a little over forty dollars a month for each member of the family; about two hundred dollars a month for a family of five. If you were eligible for food stamps, you received an extra fifty dollars or so. By the time a woman paid rent, half that money was already gone.

Thus, nearly everyone in the system had some underground source of extra money. Some women actually had jobs, doing daywork off the books. Others held part-time jobs in local stores; still others made extra cash by working policy—the Chicago numbers racket. And some women, like my mother, had surreptitious help from the men in their lives. A sporadic laborer could never be the sole support of a family. But, in combination with AFDC and food stamps, a man's paltry wage turned into enough money to move a family from desperation to subsistence.

This logic was unspoken among the women in the neighborhood, but by the time I was nine or ten, I pretty much had figured it out. The government kicked in money for the basics: rent, food, lights, gas. It was up to everybody to hustle the rest. At the time, I suspect, the government wasn't clear that it had entered into such a bargain, but we certainly were. I suspect my father felt both enraged and relieved at my mother's entry into the welfare system. It was her partial declaration of independence, and he resented it bitterly, yet we were a burden he couldn't carry along with the burden of his gambling affliction. Some part of him must have been glad to know we would not live utterly from hand to mouth.

What I'm sure he resented almost as much was my growing awareness of our situation. Even more than my smart-mouth questions, Daddy loathed my understanding of all the machinations of welfare. Once Mama and I sat at the kitchen table, fantasizing over a couch on sale at Nelson Brothers, a local furniture chain that specialized in low quality and high prices. I gazed at the price and intoned that it would cost too much—and besides, we were on welfare and the caseworker would wonder where we got it. Daddy was furious; I was confused. To keep ourselves together, we needed a common story, a common set of lies, to be told at appropriate times, a common understanding of what was at risk. There were things I had to know if I was to do my part.

Besides, I had become the mediating voice between my parents—especially my mother—and the outside world. There were always forms to fill out, or questionnaires to complete, or official-looking papers to read. Mama managed to do most of these things, but not very well. It fell to me, and sometimes to Hiawatha, to read the documents and explain what had to be done. I was the person who read the lease for our apartment before Mama signed it. I was the one who explained that the Department of Public Aid wanted verification of certain information by a certain date or our benefits would be altered, perhaps even discontinued. I did these things because I could. I did them because I was the oldest, and because it was my job to help my mother. But there was no way to do them and remain unconscious of the attitudes that surrounded welfare and the attitudes that prevailed in our household.

I was very aware, for example, of how important caseworkers were, and how important it was to give them the right impression. I didn't see them very often, usually during the summer months or during school vacations. But despite the parade of varying faces—some black, some white—over the course of my childhood, the central theme was one I copied from my mother: how to help this woman and her four children get by. My mother was easily a welfare poster child, and I knew that well enough to contribute to that image. I always made it a point to have a book in my hands when the worker came; whatever I was reading acted as a kind of cloaking device. I could be in the room, listen in, but be ignored, since I was just a little girl reading.

My mother's only bone of contention with the workers was in the matter of our schooling. More than once, a caseworker tried to impress upon my mother the inappropriateness of spending part of her monthly grant to send us all to St. Ambrose. Why couldn't she just send us to public school like everyone else? My mother was steely but gracious, as she reminded them that she wanted us to have the education she never got, an education we could not get in Chicago's segregated public schools. Only one caseworker said outright that the state of Illinois was not sending her money so that we could go to private school.

"Once you send it to me, it's not your business what I do with it," my mother answered, in a voice that brooked no further discussion. "I'm taking care of my kids."

3

Discrimination/ Racism/Stigma

As noted in Part I, the 1999 median household income for Whites was $44,366, Native Americans $30,784, Latinos $30,735, and African Americans $27,910. While the official poverty rate by race/ethnicity follows that pattern, the rates for non-Whites is more than three times the rate for Whites (the rate for Whites in 2000 was 7.5 percent, for Native Americans 25.9 percent, Latinos 21.2 percent, and African Americans 22.1 percent).

The relatively high poverty rates for African Americans, Latinos, and Native Americans is associated with various forms of institutional discrimination, such as attending underfunded schools, living in areas of concentrated poverty, being disproportionately employed in the lowest segment of the labor market where pay is at or near the minimum wage with few benefits, and relatively high unemployment. For example, in April 2001, the unemployment rate for Whites was 4.0 percent, compared with 6.5 percent for Latinos and 8.2 percent for African Americans (this African American rate of twice that for Whites has been typical for the last several decades).

Historically, African Americans have been the largest racial/ethnic minority and have experienced the greatest social and economic discrimination (George, 1999), but the demographics of minorities are changing in the United States with the recent waves of immigrants from Latin America and Asia. Data from the 2000 Census revealed that (1) for the first time Latinos outnumbered African Americans (35.3 million to 34.6 million); (2) 10.6 percent of the population was foreign born; (3) the number of Latinos increased by 60 percent from 1990 to 2000; and (4) there may be as many as 11 million undocumented people in the United States (El Nasser, 2001; Cohn, 2001). Immigrants enter at the bottom of society's stratification system, doing society's "dirty work" for low wages and few if any benefits. Moreover, because of the 1996 welfare reform passed by Congress, all legal immigrants were cut off from food stamps, and those who entered the country after the welfare bill was signed were ineligible for federal programs such as Supplemental Security Income and state-run programs such as temporary welfare and Medicaid. One consequence is the lack of health care. A study by the Center on Budget Priorities found that, nationally, 46 percent of immigrant children who were not citizens lacked health insurance in 1999, compared with 20 percent of children whose parents were native born (reported in Scher, 2001).

Closely related to discrimination is the concept of stigma. A **stigma** is an attribute that is socially devalued and disgraced. Sociologist Erving Goffman describes it this way in his classic *Stigma:*

> While the stranger is present before us, evidence can arise of his possessing an attribute that makes him different from others in the category of persons available for him to be, and of a less desirable kind—in the extreme, a person who is quite thoroughly bad, or dangerous, or weak. He is thus reduced in our minds from a whole and usual person to a tainted, discounted one. Such an attribute is a stigma, especially when its discrediting effect is very extensive; sometimes it is also called a failing, a shortcoming, a handicap. (Goffman, 1963:3)

People with stigmas have what Goffman called a spoiled identity and this spoiled identity has negative consequences:

> By definition, of course, we believe the person with a stigma is not quite human. On this assumption we exercise varieties of discrimination, through which we effectively, if often unthinkingly, reduce his life chances. We construct a stigma-theory, an ideology to explain his inferiority and account for the danger he represents, sometimes rationalizing an animosity based on other differences such as those of social class. (Goffman, 1963:5)

Goffman (1963) classifies social stigmas into three major categories. "Tribal identities" refer to stigmas involving nationality, ethnicity, race, and religion. "Abominations of the body" include physical deformities such as disabilities, physical unattractiveness, and diseases (HIV-AIDS). The final type involves stigmatizing "blemishes of individual character," such as mental disorder, criminality, drug addiction, homosexuality, and impoverishment. Various groups in society are singled out by the majority and defined as different and abnormal, justifying the majority's discrimination against those who were singled out as "inferiors." These "others" are not only different, they are also deficient. Thus, their identity is negatively evaluated and stereotyped—i.e., stigmatized.

THE STIGMATIZING OF THE IMPOVERISHED

According to a survey conducted in early 2001, about half of Americans believe that poverty is a combination of biological and cultural factors (National Public Radio, 2001). In effect, many Americans believe that poverty is a self-inflicted condition resulting from personal factors such as a lack of effort and/or ability (Chafel, 1997). The common perception is that the poor are not only lazy but also dependent on government welfare—characteristics that Americans despise. Young, poor women are especially stereotyped, stigmatized, and demonized. They are reviled for having children out of wedlock (a "culture of illegitimacy") and for transmitting negative values; their single-parent family structure is blamed for fostering violence, crime, failure in school, and for passing their poverty on to the next generation—all traits fitting with the culture of poverty (Sidel, 1996:167).

> Poor, single mothers . . . are being portrayed as the ultimate outsiders—marginalized as nonworkers in a society that claims belief in the work ethic, marginalized as single parents in a society that holds the two-parent, heterosexual family as the desired norm, and marginalized as poor people in a society that worships success and material rewards. (Sidel, 1996:6-7)

THE CONSEQUENCES OF STIGMATIZING THE IMPOVERISHED

The foremost consequence of stigmatizing the impoverished—when most of society rejects the poor, believing them to be not only different but also inferior or, in Goffman's words, "not quite human"—is that many of the poor internalize these negative stereotypes, thus feeling shame, humiliation, and disgrace. In effect, they accept society's definition and blame themselves for their failure, resulting in a self-fulfilling prophecy.

There are several other related consequences of demonizing the poor, especially poor mothers. The poor are viewed as deserving their fate because it is their own fault. Thus, the poor are seen as undeserving of government programs such as Aid to Families with Dependent Children (AFDC), food stamps, housing subsidies, and Head Start for their children. As David Boaz, executive vice-president of the Cato Institute, a libertarian organization, put it:

> We've made it possible for a teenage girl to survive with no husband and no job. . . . If we had more stigma and lower benefits, might we end up with 100,000 bastards every year rather than a million children born to alternative families? (quoted in Sidel, 1996:6)

This type of logic justified the 1996 welfare legislation that dismantled AFDC, which is discussed in Part IV of this book. This "solution" to the problem of poverty assumed that the reasons for poverty are within the poor, not in the structure of society. In this way, society is absolved of blame, and solutions are aimed at changing poor people rather than changing society. Finally, the stereotyping and stigmatizing of the poor (or any other minority group) assumes that the poor are a homogeneous group, when they are actually heterogeneous, differing widely in personal characteristics and circumstances.

The voices in this chapter reveal the discrimination and racism experienced by welfare mothers, those who live in trailers, migrant workers, and African Americans in U.S. society.

NOTES AND SUGGESTIONS FOR FURTHER READING

Bray, Rosemary L. (1998). *Unafraid of the Dark*. New York: Random House.

Brunious, Loretta J. (1998). *How Disadvantaged Adolescents Socially Construct Reality: Do You Hear What I Hear?* New York: Garland.

Chafel, Judith A. (1997). "Societal Images of Poverty." *Youth and Society* 28 (June):432–463.

Cohn, D'Vera (2001). "Another Case of an Undercount." *Washington Post National Weekly Edition* (March 26):31.

El Nasser, Haya (2001). "Census Shows Greater Numbers of Hispanics." *USA Today* (March 8):3A.

Fair, Bryan K. (1997). *Notes of a Racial Caste Baby: Color Blindness and the End of Affirmative Action*. New York: New York University Press.

George, Hermon, Jr. (1999). "Black America, the 'Underclass,' and the Subordination Process." Pp. 197–213 in *A New Introduction to Poverty: The Role of Race, Power, and Politics,* Louis Kushnick and James Jennings (eds.). New York: New York University Press.

Goffman, Erving (1963). *Stigma: Notes on the Management of Spoiled Identity*. Englewood Cliffs, NJ: Prentice-Hall.

Harrington, Michael (1963). *The Other America: Poverty in the United States*. Baltimore: Penguin Books.

Hartman, Chester W., Bill Bradley, and Julian Bond (1997). *Double Exposure: Poverty and Race in America*. New York: M. E. Sharpe.

Heatherton, Todd F., Robert E. Kleck, Michelle R. Hebl, and Jay G. Hull (eds.), (2000). *The Social Psychology of Stigma*. New York: The Guilford Press.

hooks, bell (2000). *Where We Stand: Class Matters*. New York: Routledge.

Katz, Jane (ed.), (1995). *Messengers of the Wind: Native American Women Tell Their Stories*. New York: Ballantine Books.

National Public Radio (2001). "Poverty in America." <http://search1.npr.org/search97cgi/>

Olinger, David, and Jeffrey A. Roberts (2001). "Loans Cost Minorities More." *Denver Post* (February 27):1A, 10A.

Quadagno, Jill (1994). *The Color of Welfare: How Racism Undermined the War on Poverty*. New York: Oxford University Press.

Scher, Abby (2001). "Access Denied: Immigrants and Health Care." *Dollars & Sense* #235 (May/June):8.

Schiller, Bradley R. (1995). *The Economics of Poverty & Discrimination*. Englewood Cliffs, NJ: Prentice-Hall.

Sennett, Richard, and Jonathan Cobb (1973). *The Hidden Injuries of Class*. New York: Random House Vintage Books.

Sidel, Ruth (1996). *Keeping Women and Children Last: America's War on the Poor.* New York: Penguin.

Waxman, Chaim I. (1977). *The Stigma of Poverty: A Critque of Poverty Theories and Policies*. New York: Pergamon Press.

Wray, Matt, and Annalee Newitz (eds.), (1997). *White Trash: Race and Class in America*. New York: Routledge.

Zuccino, David (1997). *Myth of the Welfare Queen*. New York: Scribner's.

My Name Is Not "Those People"

Julia Dinsmore

As stated in the introduction, the relatively high frequency of divorce and the large number of never-married women with children have resulted in high numbers of poor, single-female-headed families (27.8 percent of single-headed families live in poverty, compared with 4.8 percent of two-parent families). The following selection shows the anger of one single mother on welfare. It was written for the Sept/Oct 1998 Newsletter of People United For Families (PUFF), a Denver-based grassroots organization of low-income individuals and families. Her words reflect some of the common misconceptions and stereotypes about women on welfare. Which theoretical perspective is reflected in her words? Think about what she means when she says, "For I am not the problem, but the solution."

My name is not "Those People." I am a loving woman, a mother in pain, giving birth to the future, where my babies have the same chance to thrive as anyone.

My name is not "Inadequate." I did not make my husband leave us—he chose to, and chooses not to pay child support. Truth though, there isn't a job for all Fathers to support their families.

My name is not "Problem and Case to Be Managed." I am a capable human being and citizen, not a client. The social service system can never replace the compassion and concern of loving Grandparents, Aunts, Uncles, Fathers, Cousins, Community—all the bonded people who need to be but are not present to bring the children forward to their potential.

SOURCE: Julia Dinsmore, "My Name is Not 'Those People,'" People United For Families (PUFF) Poverty to Prosperity Newsletter (September/October 1998). Reprinted by permission of the publisher.

My name is not "Lazy, Dependent Welfare Mother." If the unwaged work of parenting, homemaking and community building was factored into the Gross National Product, my work would have untold value. And I wonder why middle-class sisters whose husbands support them to raise their children are glorified and they don't get called lazy and dependent.

My name is not "Ignorant, Dumb or Uneducated." I live with an income of $621 with $169 in food stamps. Rent is $585. That leaves $36 a month to live on. I am such a genius at surviving that I could balance the state budget in an hour.

Nevermind that there is a lack of living-wage jobs. Nevermind that it is impossible to be the sole emotional, social and economic support to a family. Nevermind that parents are losing their children to the gangs, drugs, stealing, prostitution, social workers, kidnapping, the streets, the predator. Forget about putting money into our schools—just to build more prisons.

My name is not "Lay Down and Die Quietly." My love is powerful and my urge to keep my children alive will never stop. All children need homes and people who love them. They need safety and the chance to be the people they were born to be.

The wind will stop before I let my children become a statistic. Before you will give in to the urge to blame me, the blame that lets us go blind and unknowing into the isolation that disconnects us, take another look. Don't go away. For I am not the problem, but the solution. And . . . My name is not "Those People."

The Dynamics of Welfare Stigma

Robin Rogers-Dillon

In this selection, Assistant Professor of Sociology at Queens College Robin Rogers-Dillon interviews ten divorced/separated women who use public assistance. They are White, Latina, and African American, and they came from different social class backgrounds before their divorce/separation. As you read this selection, consider the following questions: What is the difference between those recipients who came from middle-class families and those who came from families that had received public assistance at some time? What does this difference say about the culture of poverty argument (regarding the cyclical nature of poverty)? What were some of the strategies women used to avoid the stigma of food stamps? Can you think of a possible alternative to food stamps, something that could help lessen the stigma of using them?

THE EXPERIENCE OF WELFARE STIGMA

The women in this study described going on to welfare in terms of necessity. The stigma of receiving welfare was almost meaningless in the face of pressing needs for food, shelter, and other goods. Most of the respondents reported that going on to welfare was not a difficult decision to make, with no job or child support, they had no other options.

> It is instinctive. You don't have to think about there is a crumb and I'm hungry, go get it. It's survival. You do what you have to do. It is demeaning. I hate it. (Diane, white, poor during marriage)

> No other options. No more money could I get, (. . .) no more things could I sell. I didn't have anything left to sell, so it is either welfare or no fare, you know? (Erica, African American, middle class during marriage)

Though the stigma of welfare was not a deterrent to applying for assistance, respondents were very aware of the stereotypes of welfare recipients. They saw the public's image of most welfare recipients as one of lazy, baby-making women living off of other people's labor.

> Well [people think] that's another welfare case, you know. She's lazy. Basically that's what I hear. I hear a lot that you know, they're lazy, they don't wanna do nothing and all this kind of stuff, (. . .) A lot of men have like a stigma about women being on DPA. And a lot of them think that, you know, you lazy and you not trying to do better, all that kind of stuff. Then [I] feel a lot of times I wouldn't even let them know. (Jenny, African American, working class during marriage)

> I mean when you're on welfare you really supposed to be looking for a man, you know? It's like come and take me. (. . .) Well, [people think] you're staying home all day making more babies. (Erica, African American, middle class during marriage)

Kim, the 18-year-old daughter of a respondent, spoke about her own feelings

SOURCE: Robin Rogers-Dillon, "The Dynamics of Welfare Stigma," *Readings in Sociology,* 3rd ed., Garth Massey (ed.) (New York: W. W. Norton, 2000), pp. 255–263. By permission of the author.

towards public assistance recipients, before her mother became one:

> I think if someone had said they were on welfare to me, my immediate reaction would have been to the people, they are lazy and they are unmotivated and they are not as smart and they didn't get a good education and they are bleeding the government and they are bleeding us. (Kim, white, formerly middle class)

Kim's description illustrates how powerfully negative the image of welfare recipients can be.

Many of the respondents felt the need to defend themselves against unflattering assumptions that they felt others made about them based on their status as welfare recipients. For example, Erica, who only has one child, feels that by virtue of being on welfare she must defend herself against accusations that women on welfare have many children. She feels that she must distance herself from this image of the "welfare mother" even though it does not accurately reflect her situation.

> To be honest, I'm going to tell you, frankly as far as men feel about it, men feel women on welfare have a lot of children. They are promiscuous. As far as being in the African American community, it's like if you have a lot of children, forget it. Because it's like everybody is looking for a dad. And I'm not looking for a damn daddy. (Erica, African American, middle class during marriage)

The stereotype of the "welfare mother" exists apart from the actual situations of the women in this study and each of the women had to face this stereotype, but they did so in different contexts. The meaning of welfare was different for each woman, and it was shaped by her life history. For example, for Lynn, a young woman who had only a brief marriage and knew a number of other women on public assistance, welfare held little symbolic meaning, and it gave her the opportunity to go back to school.

> You know, I was like I was supposed to get this degree. . . . Not that I don't wanna get off [welfare] and be independent. (Lynn, African American, poor during marriage)

In contrast, for Lisa, a middle-aged woman who had never been poor before her divorce, food stamps symbolized her fall from the middle class and were imbued with the pain of having lost her marriage, middle class standing and the future for which she had worked for twenty years. Lisa describes trying to make sense of her unexpected entrance into the welfare system in the following way:

> I think it's [her feelings of humiliation over receiving welfare] probably a reaction to where I am now, which is not at all where I expected to be. It's interesting, like the first time I went in there, I was talking to the first in-take person, I just broke down and sobbed. I said I can't believe I'm here. I can't believe it. And she said to me. . . . She was actually very nice about it. She said to me, "Lisa, you don't understand." She said, "More and more over the last five years we've had more women who were wives or ex-wives of doctors and lawyers and very successful businessmen come in and say those exact words." You just cannot believe that, I guess, that your life has taken such a major turn. (Lisa, white, middle class during marriage)

Being on welfare simply does not make sense to Lisa. Similarly, Erica, an African American woman who had surpassed her parents' economic position during her marriage, now bitterly compares her position to theirs:

> . . . and it was like my parents are so successful and not having any damn problems and that was a big thing, that it was just me. It was a total admission of failure, you know? (Erica,

> African American, middle class during marriage)

Another formerly middle class woman, Amy, who recently separated from her husband, explained to me how welfare benefits should be used. In the course of her explanation, it became evident that Amy was struggling to come to terms with her new economic position. Amy sounded confident and assertive as she discussed what welfare recipients should do in the abstract, but toward the end of her statement Amy's focus shifted to her own situation and her voice trailed off until it was almost inaudible:

> I think that anyone who is eligible [for welfare benefits] should be able to receive it if they chose to, but I don't think they—that you should look at that [welfare] as something that you are always going to get. You should always look to the future and see that you've got to get off it. This is just, for me this is just kind of a stopgap thing to get me through where I can afford . . . something. . . . I don't know what . . . just to manage. (Amy, white, middle class during marriage)

The fall from the middle class created tremendous confusion and anxiety for Linda, Erica and Amy. Respondents who came from families that had received public assistance at some time, on the other hand, did not experience the same kind of confusion. This is not to say that being on welfare was necessarily any easier, or even more acceptable, for them. For example, Diane, who had a very difficult relationship with her mother, a welfare recipient herself, saw receiving welfare as being a part of a painful legacy:

> She [her mother] probably expected me to continue the legacy of raising children on welfare and doing nothing with my life. If I did anything other than that she would probably be very disappointed with herself. (Diane, white, poor during marriage)

In contrast, Sally, whose mother had also been on welfare, was able to use her mother's experiences, and those of other family members who had received welfare, to legitimate her own use of welfare:

> Public assistance is not a taboo. I mean my mother was on it when she needed to be, when she was separated from my father and she had lost her two businesses she went on public assistance. She went to college when she was on public assistance and once she got a job she got off of it. So, that's the way I saw it. You use it until, you know, you have an income. (Sally, Latina, poor during marriage)

As the last two examples illustrate, former class status, while important, did not define how respondents experienced being on welfare. Rather, a combination of the respondent's social locations and personal histories seem to have shaped their feelings about being on welfare. The specific ways in which class, race and personal histories influenced the respondents' construction of personal identity is beyond the scope of this paper. The findings in this study, however, do suggest that the effects of personal background on welfare recipients' experiences of stigma and construction of personal identities may be fertile ground for further research.*

FOOD STAMPS AND STIGMA MANAGEMENT

Food stamps are highly visible stigma symbols, they instantly, and very publicly, reveal the user's economic position and status as a welfare recipient. Even

*See David Snow and Leon Anderson's *Down on Their Luck* (1993), especially chapter 7, for a discussion of some of the ways homeless street people construct and negotiate their personal identities in light of their stigmatized social identities.

respondents who stated that receiving welfare was nothing to be ashamed of bristled over having their status as public assistance recipients revealed in the course of their daily activities and without their consent. The women often talked about having lost their privacy and about being frustrated that other people "know their business." For example, Diane explained why she dislikes the objects that mark her as a welfare recipient:

> Food stamps and that little stupid blue health pass card they give you is just another way of labeling you. Your status is reduced now. You're not who you walked in as because now they see that you're less than in their mind, you're a welfare recipient. (Diane, white, poor during marriage)

Being labeled as a welfare recipient is extremely painful for Diane, and she generally tries to keep her status as a welfare recipient a secret. But this is not always possible, as the following story illustrates.

> When I moved here, I figured a new neighborhood, nobody knows my business. . . . The building I moved up out [of] in West Philly, everybody was on welfare. . . . [I]t was that kind of building. Everybody had food stamps and money problems and the whole bit. When I moved here I thought, oh good, this . . . I mean there's a lot of poor people but they don't know my business. They don't know me and I can come in with some anonymity and it didn't work, because my landlord told everybody, oh the girl that's moving in she's on Section 8. He lives six houses down. So everybody knew my business and I felt so like . . . With my mask and I couldn't even put it on, you know? (Diane, white, poor during marriage)

It is particularly hard for Diane to avoid being revealed as a welfare recipient when she uses food stamps:

> One day I was in the store and the lady needed change. You have to pay five. . . . There's five, tens and ones. And she said, I need change. I need some ones. And the girl several registers down said, I just gave you ones. And she said, For *food stamps!* I couldn't crawl small enough and close enough to the ground I felt so bad. (Diane, white, poor during marriage)

Many of the respondents described feeling awkward using food stamps in front of particular people or groups of people. For example, Gina, who was raised in a working class family, talked about how she feels more hesitant to use food stamps in white neighborhoods. In her description, the social audience and situation in grocery stores outside of the inner city create a stigma that she does not feel when she shops in inner city neighborhoods:

> Sometime I go and I don't shop in, say the inner city neighborhoods, my girlfriend and I we'll get in her car and we'll drive out to say a white neighborhood to go shopping—Super Fresh, or you know things like that and sometimes you are afraid, you don't want other people to see you pulling out food stamps. (Gina, African American, poor during marriage)

Another respondent, Lisa, avoids being seen by people who know her by shopping in a grocery store outside of her neighborhood:

> When I know I have to go shopping now I go to another store altogether. . . . [N]ormally I would go to the Acme here. I've been going there for 16 years. Most of the people, the checkers, know me. Most of them have seen my kids grow up. Whereas if I go to the other one they don't know me and I am just kind of a face in the crowd. I don't know how to put this equation together. If you

> know what I mean. And I don't know whether it's something I perceive or if it's wished that when I pay with food stamps it's almost like I'm a nonentity. But I don't know if they're really reacting to me that way or if I choose to. . . . They sit there looking at me that way. (Lisa, white, middle class during marriage)

Several of the women in this study were also concerned with how their children managed their use of food stamps. Pam, a young woman who had been raised in a working class family and was homeless for a time after escaping an extremely violent marriage, explains that she tries to teach her son to be discreet when he uses food stamps:

> My oldest son, with the food stamps, he thinks it's like hey! And I told him, this is nothing you wanna, he wants to get food stamps and go to the corner store and buy some chips, a bag of chips and a soda. And I tell him no, it's not a thing like that. . . . It is like our family secret. . . . They can remember when we was in the shelter and so I tell them this is just like that. We just have to go through, but it's not something we want to make flyers and put out and make everybody know. (Pam, African American, poor during marriage)

In contrast, Jenny, who is also from a working class background, has tried to teach her daughter not to feel awkward about using food stamps:

> So I learned in fact she [her daughter] felt funny about it especially when I would send her to the store with food stamps or whatever. You know, I didn't know it but she told me she was ashamed to use them. I thought the kid was just being lazy about going to the store. . . . But when she found out some kind of way that some of her friends were on DPA—so like now she is starting to understand that we're not the only ones in this boat, you know. (Jenny, African American, working class during marriage)

Though Pam and Jenny view food stamps differently, they have both had to teach their children how to manage food stamps and the information that food stamps convey.

Food stamps convey a degraded social status, and yet they also provide essential goods. Respondents disliked the social meaning of food stamps, but most also found food stamps helpful for budgeting and getting through the month. When I asked the women if they thought that food stamps should be replaced with cash, almost all said no. For the respondents, balancing the practical advantages of food stamps with potentially negative social consequences was a central task.

The Avoidance Strategy

Because of their power to discredit, almost all of the women in this study could think of times when they had chosen not to use food stamps. Using cash rather than food stamps was a common strategy used by respondents to avoid being labeled as welfare recipients. For example, Pam reported:

> One time I was with a friend, it was as guy, not my boyfriend but it was a friend and we was in the Acme and I felt so bad afterward because I only had a couple of dollars, you know green dollars, but I felt the way that I didn't want him to know that I was receiving public assistance. (Pam, African American, poor during marriage)

Similarly, Sally, a young woman whose mother had also received welfare, reported that she had chosen not to use food stamps when she recognized a person in the grocery store:

> I think someone was in line that was a parent of my son's [friend] . . . something like that, and I don't like them knowing my business. (Sally, Latina, poor during marriage)

Diane chose not to use food stamps to preserve an image of herself in front of two men that she did not even know:

> And then one time I was shopping and there were two good looking, really good looking guys—a couple of years ago when I was really thin. And they were making little comments about me. And I was going on a picnic with my then best friend Judy, who was the one on welfare, single parent and section 8 too. And I wouldn't pay with my food stamps. I paid in cash because I didn't want them to get a different idea of me. And she got on my case all the way to the car, all the way to the park. She said, that was so stupid. Why would you do that? I said, Judy, everybody was looking at me. They noticed me and I didn't want them to think any less of me, so, you know, I paid cash instead of stamps. Now, I know I'm going to suffer at the end of the month for doing that, but I did it. (Diane, white, poor during marriage)

Terri, who lives in an area where food stamps are common, often avoids being publicly discredited altogether by sending her boyfriend to the grocery store with her food stamps rather than going herself:

> I don't go to the store. I send my boyfriend. I call him and he says "I don't care [about using food stamps], I'll go." Then he goes to the store. (Terri, Latina, poor during marriage)

Using food stamps in the grocery store brought the women's status as welfare recipients into sharp focus. In front of specific social audiences, such as boyfriends, neighbors, et cetera, being labeled as a welfare recipient was particularly troublesome to the respondents. In order to avoid being discredited in these situations, respondents employed several strategies: using cash rather than food stamps, shopping in different neighborhoods, and sending someone else to the grocery store. These strategies, especially using cash rather than food stamps, are difficult to sustain. The respondents had to choose carefully in which situations being discredited outweighed the relative hardships of an avoidance strategy. The calculations made by respondents highlights the dynamic nature of welfare stigma; if welfare stigma were static, then there would be no need to employ strategies that can only be used sporadically.

The Recasting Strategy

Recasting welfare in a more positive light was another strategy used by respondents to manage welfare stigma. Though they could not eliminate the negative public images of welfare recipients, respondents could, and did, recast welfare in a more positive light for themselves. Knowing that other people used or accepted food stamps helped some of the respondents to view their own use of food stamps more positively. In a striking example of recasting welfare, Lynn "discovered" (incorrectly) that food stamps were legal tender outside of the United States:

> And as I found out from my sister, she is really intelligent. She's been around and everything. She told me she's taken her food stamps outside the United States and used them because you can. . . . She said they are legal tender just like money. I was shocked, you know. She was like with somebody important. And she went to the store and pulled out food stamps. I was like, oh my god. I was like, you go girl. I didn't know that. . . . It's just to know that . . . to know that they're legal tender elsewhere. (Lynn, African American, poor during marriage)

Hearing that her sister was able to use food stamps outside of the United States helped Lynn to view food stamps as being legitimate, "just like money" in every way. Similarly, seeing her sister use food stamps in front of "someone important" helped Lynn to be less ashamed about using food stamps herself. In a similar story, Jenny, an African American respondent, explained that seeing white people use food stamps helped her to view food stamps as being legitimate and nothing to be embarrassed about:

> I was pretty much embarrassed. . . . Then I seen some of the white people around here with them, I said, "oh well that's it." So now it didn't bother me. (Jenny, African American, working class during marriage)

Sally, a young Latina woman, was able to construct a positive image of welfare recipients based on her own experiences:

> The only images [of welfare recipients] I had were of my mother or my family members and I knew that they were on the system because they needed to be on the system. (Sally, Latina, poor during marriage)

The positive image of welfare recipients that Sally gained from her family helped her to legitimate being on welfare herself.

Lisa and Erica, two formerly middle class respondents, framed their being on welfare as a consequence of larger social forces. Lisa was able to depersonalize and manage the stigma of welfare by seeing her economic decline as being, in part, a result of an economic recession.

> There's so many people in unemployment and then on the welfare when unemployment runs out. Many of them have no alternative but to go on welfare whether it's food stamps or, you know, total assistance. And again that's not through any fault of their own. And I don't know what the proportions are, but my guess is there's probably a whole lot bigger numbers of people on some form of public assistance now as a direct result. I mean, I call it a depression. (Lisa, white, middle class during marriage)

Lisa also viewed being on welfare as the result of her divorce and of unjust divorce laws:

> I think with the divorce rate coming up, a lot of women . . . the divorce laws totally inadequate, totally inadequate. Leaves a lot of women who otherwise would still be in whatever marriage. Leaves them out on a limb because the laws aren't there to protect their interests and their children's interest. (Lisa, white, middle class during marriage)

Erica viewed her entrance onto welfare as the combined result of her divorce, economic hard times and racism. During her marriage Erica had achieved a middle class household income of around $40,000. After her husband left, Erica struggled financially but survived. She supported herself and her son for several years before she lost her job when the hospital at which she worked was sold:

> I'm going to be honest and tell you. Being African American, there's a lot of us who are out of work and there is a lot of us who are. . . . If we're getting the assistance ourselves [or] we know somebody who say getting assistance and it is not the big embarrassment that it might be, other people, you know. And you're the last hired, first fired. You tend to have to find stopgap measures. (Erica, African American, middle class during marriage)

One interesting aspect of the recasting strategy is that there seemed to be a class pattern to it. Formerly working and lower class respondents, such as Lynn, Sally, and Jenny, tended to recast welfare itself in a positive light, seeing it as a legitimate form of income. In contrast, formerly middle class respondents, such as Lynn and Erica, emphasized the larger social forces, such as economic hard times combined with racial and gender inequality, that made their receipt of welfare inevitable. Seeing their being on welfare as the inevitable outcome of social inequality, the middle class women were able to deflect some of the stigma of welfare. In both of these strategies, the respondents redefined the meaning of their being on welfare in a more positive light.

Sunset Trailer Park

Allan Bérubé with Florence Bérubé

Many have heard of the term "trailer trash" and the stereotypes that go along with that term. On the Internet, one can find thousands of websites and even items for sale under the term "trailer trash." Thinking sociologically, where do these stereotypes come from, and how do they affect the people that they supposedly "describe"? In this essay, Allan Bérubé talks with his mother about growing up as part of a white, working-class family in the Sunset Trailer Park in Bayonne, New Jersey. In his essay, Bérubé talks about the two conflicting stereotypes of trailer park life in the popular culture of the 1950s. Do these same stereotypes exist today? What type of social hierarchy existed within the park? Do these hierarchies exist in all types of neighborhoods? Do people "naturally" rank each other, regardless of social class? What are the inevitable effects of such hierarchies?

"I cried," my mother tells me, "when we first drove into that trailer park and I saw where we were going to live." Recently, in long-distance phone calls, my mother —Florence Bérubé—and I have been digging up memories, piecing together our own personal and family histories. Trailer parks come up a lot.

During the year when I was born—1946—the booming, postwar "trailer coach" industry actively promoted house trailers in magazine ads like this one from the *Saturday Evening Post*:

> *Trailer Coaches Relieve Small-Home Shortage Throughout the House-Hungry Nation*
>
> Reports from towns and cities all over the United States show that modern, comfortable trailer coaches—economical and efficient beyond even the dreams of a few years ago—are playing a major part in easing the need for small-family dwellings. Returning veterans (as students or workers), newlyweds, and all others who are not ready for—or can't locate—permanent housing, find in the modern trailer coach a completely furnished (and amazingly comfortable) home that offers the privacy and efficiency of an apartment coupled with the mobility of an automobile.

When I was seven, my parents, with two young children in tow, moved us all into a house trailer, hoping to find the comfort, privacy, and efficiency that the "trailer coach" industry had promised. But real life, as we soon discovered, did not imitate the worlds we learned to desire from magazine ads.

Dad discovered "Sunset Trailer Park" on his own and rented a space for us there before Mom was able to see it. On

SOURCE: Allan Bérubé with Florence Bérubé, "Sunset Trailer Park," in *White Trash*, Matt Wray and Annalee Nevitz (eds.) (New York: Routledge, 1997), pp. 15–20. Reproduced by permission of Routledge, Inc., part of the Taylor & Francis Group.

our moving day in January 1954, we all climbed into our '48 Chevy and followed a rented truck as it slowly pulled our house trailer from the "Sunnyside Trailer Park" in Shelton, Connecticut, where we lived for a few months, into New York State and across Manhattan, over the George Washington Bridge into New Jersey, through the garbage incinerator landscape and stinky air of Secaucus—not a good sign—then finally into the Sunset Trailer Park in Bayonne.

A blue-collar town surrounded by Jersey City, Elizabeth and Staten Island, Bayonne was known for its oil refineries, tanker piers and Navy yard. It was a small, stable, predominantly Catholic city of working-class and military families, mostly white with a small population of African Americans. When we moved there, Bayonne was already the butt of jokes about "armpits" of the industrial Northeast. Even the characters on the TV sitcom *The Honeymooners,* living in their blue-collar world in Brooklyn, could get an easy laugh by referring to Bayonne. "Ralph," Alice Kramden says to her husband in one episode, "you losing a pound is like Bayonne losing a mosquito." My mom was from Brooklyn, too. A Bayonne trailer park was not where she wanted to live or raise her children.

Along with so many other white working-class families living in fifties trailer parks, my parents believed that they were just passing through. They were headed toward a *Better Homes and Gardens* suburban world that would be theirs if they worked hard enough. We moved to Bayonne to be closer to Manhattan, where Dad was employed as a cameraman for NBC. He and his fellow TV crewmen enjoyed the security of unionized, wage-labor jobs in this newly expanding media industry. But they didn't get the income that people imagined went with the status of TV jobs. Dad had to work overtime nights and on weekends to make ends meet. His dream was to own his own house and start his own business, then put us kids through school so we would be better off and not have to struggle so much to get by. My parents were using the cheapness of trailer park life as a stepping-stone toward making that dream real.

As Dad's job and commuting took over his life, the trailer park took over ours. We lived in our trailer from the summer of 1953 through December 1957, most of my grade school years. And so I grew up a trailer park kid.

Sunset Trailer Park seemed to be on the edge of everything. Bayonne itself is a kind of land's end. It's a peninsula that ends at New York Bay, Kill Van Kull and Newark Bay—polluted bodies of water that all drain into the Atlantic Ocean. You reached our trailer park by going west to the very end of 24th Street, then past the last house into a driveway where the trailer lots began. If you followed the driveway to its end, you'd stop right at the waterfront. The last trailer lots were built on top of a seawall secured by pilings. Standing there and looking out over Newark Bay, you'd see tugboats hauling barges over the oil-slicked water, oil tankers and freighters carrying their cargoes, and planes (no jets yet) flying in and out of Newark Airport. On hot summer nights the steady din of planes, boats, trucks and freight trains filled the air. So did the fumes they exhaled, which, when mixed with the incinerator smoke and oil refinery gasses, formed a foul atmospheric concoction that became world-famous for its unforgettable stench.

The ground at the seawall could barely be called solid earth. The owner of the trailer park occasionally bought an old barge, then hired a tugboat to haul it right up to the park's outer edge, sank it with dump-truck loads of landfill, paved it over with asphalt, painted white lines on it and voila!—several new trailer lots were available for rent. Sometimes the ground beneath these new lots would sink, so the trailers would have to be moved way until the sinkholes were filled in. Trailers parked on lots built over rotten barges along the waterfront—this was life on a geographic edge.

It was life on a social edge, too—a borderland where respectable and "trashy" got confused.

* * *

"Did you ever experience other people looking down on us because we lived in a trailer park?" I ask my mom.

"Never," she tells me.

"But who were your friends?"

"They all lived in the trailer park."

"What about the neighbors who lived in houses up the street?"

"Oh, they didn't like us at all," she says. "They thought people who lived in trailers were all low-life and trash. They didn't really associate with us."

In the 1950s, trailer parks were crossroads where the paths of poor, working-class and lower-middle-class white migrants intersected as we temporarily occupied the same racially segregated space—a kind of residential parking lot—on our way somewhere else. Class tensions—often hidden—structured our daily lives as we tried to position ourselves as far as we could from the bottom. White working-class families who owned or lived in houses could raise their own class standing within whiteness by showing how they were better off than the white residents of trailer parks. We often responded to them by displaying our own respectability and distancing ourselves from those trailer park residents who were more "lower-class" than we were. If we failed and fell to the bottom, we were in danger of also losing, in the eyes of other white people, our own claims to the racial privileges that came with being accepted as white Americans.

In our attempt to scramble "up" into the middle class, we had at our disposal two conflicting stereotypes of trailer park life that in the 1950s circulated through popular culture. The respectable stereotype portrayed residents of house trailers as white World War II veterans, many of them attending college on GI loans, who lived with their young families near campuses during the postwar housing shortage. In the following decades, this image expanded to include the predominantly white retirement communities located in Florida and the Southwest. In these places, trailers were renamed "mobile" or "manufactured" homes. When parked together, they formed private worlds where white newlyweds, nuclear families and retirees lived in clean, safe, managed communities. You can catch a glimpse of this world in the 1954 Hollywood film *The Long, Long Trailer,* in which Lucille Ball and Desi Arnaz spend their slapstick honeymoon hauling a house trailer cross country and end up in a respectable trailer park. (The fact that Arnaz is Cuban-American doesn't seriously disrupt the whiteness of their Technicolor world—he's assimilated as a generically "Spanish" entertainer, an ethnic individual who has no connection with his Cuban American family or community).

A conflicting stereotype portrayed trailer parks as trashy slums for white transients—single men drifting from job to job, mothers on welfare, children with no adult supervision. Their inhabitants supposedly engaged in prostitution and extramarital sex, drank a lot, used drugs, and were the perpetrators or victims of domestic violence. With this image in mind, cities and suburbs passed zoning laws restricting trailer parks to the "other side of the tracks" or banned them altogether. In the fifties, you could see this "white trash" image in B-movies and on the covers of pulp magazines and paperback books. The front cover of one "trash" paperback, "Trailer Park Woman," proclaims that it's "A bold, savage novel of life and love in the trailer camps on the edge of town." The back cover, subtitled "Temptation Wheels," explains why trailers are the theme of this book.

> Today nearly one couple in ten lives in a mobile home—one of those trailers you see bunched up in cozy camps near every sizable town. Some critics argue that in such surroundings love tends to become casual. Feverish affairs take place virtually right out in the open. Social codes take strange and shocking twists . . . 'Trailer Tramp' was what they called Ann Mitchell—for she symbolized the twisted morality of the trailer camps . . . this book shocks not by its portrayal of her degradation—rather, by boldly bringing to light the conditions typical of trailer life.

This image has been kept alive as parody in John Waters' independent films, as reality in Hollywood films such as *Lethal Weapon, The Client* and *My Own Private Idaho,* and as retro-fifties camp in contemporary postcards, posters, t-shirts and refrigerator magnets.

I imagine that some fifties trailer parks did fit this trashy stereotype. But Sunset Trailer Park in Bayonne was respectable—at least to those of us who lived there. Within that respectability, however, we had our own social hierarchy. Even today, trying to position ourselves into it is difficult. "You can't say we were rich," as my mom tries to explain, "but you can't say we were at the bottom, either." What confused things even more were the many standards by which our ranking could be measured—trailer size and model, lot size and location, how you kept up your yard, type of car, jobs and occupations, income, number of kids, whether mothers worked as homemakers or outside the home for wages. Establishing where you were on the trailer park's social ladder depended on where you were standing and which direction you were looking at any given time.

To some outsiders, our trailer park did seem low-class. Our neighbors up the street looked down on us because they lived in two- or three-family houses with yards in front and back. Our trailers were small, as were our lots, some right on the stinky bay. The people in houses were stable; we were transients. And they used to complain that we didn't pay property taxes on our trailers, but still sent our kids to their public schools.

On our side, we identified as "homeowners" too (if you ignored the fact that we rented our lots) while some people up the street were renters. We did pay taxes, if only through our rent checks. And we shared with them our assumed privileges of whiteness—theirs mostly Italian, Irish, and Polish Catholic, ours a more varied mix that included Protestants. The trailer park owner didn't rent to Black families, so we were granted the additional status of having our whiteness protected on his private property.

The owner did rent to one Chinese American family, the Wongs (not their real name), who ran a Chinese restaurant. Like Desi Arnaz, the presence of only one Chinese family didn't seriously disrupt the dominant whiteness of our trailer park. They became our close friends as we discovered that we were almost parallel families—both had the same number of children and Mrs. Wong and my mother shared the same first name. But there were significant differences. Mom tells me that the Wongs had no trouble as Asian-Americans in the trailer park, only when they went out to buy a house. "You don't realize how discriminatory they are in this area," Mrs. Wong told my mother one day over tea. "The real estate agents find a place for us, but the sellers back out when they see who we are." Our trailer park may have been one of the few places that accepted them in Bayonne. They fit in with us because they, unlike a poorer family might have been, were considered "respectable." With their large trailer and their own small business, they represented to my father the success he himself hoped to achieve someday.

While outsiders looked down on us as trailer park transients, we had our own internal social divisions. As residents we did share the same laundry room, recreation hall, address and sandbox. But the owner segregated us into two sections of his property: Left courtyard for families with children, right courtyard for adults, mostly newlyweds or retired couples. In the middle were a few extra spaces where tourists parked their vacation trailers overnight. Kids were not allowed to play in the adult section. It had bigger lots and was surrounded by a fence, so it had an exclusive air about it.

The family section was wilder, noisier and more crowded because every trailer had kids. It was hard to keep track of us, especially during summer vacation. Without having to draw on those who lived in the houses, we organized large group games—like Red Rover and bicycle circus shows—on the common asphalt driveway. Our activities even lured some kids away from the houses into the trailer park, tempting them to defy their parents' disdain for us.

We defended ourselves from outsiders' stereotypes of us as low-life and weird by increasing our own investment in respectability. Trashy white people lived somewhere else—probably in other trailer parks. We could criticize and look down on them, yet without them we would have been the white people on the bottom. "Respectable" meant identifying not with them, but with people just like us or better than us, especially families who owned real houses in the suburbs.

My mom still portrays our lives in Bayonne as solidly middle-class. I'm intrigued by how she constructed that identity out of a trailer park enclave confined to the polluted waterfront area of an industrial blue-collar town.

"Who were your friends?" I ask her.

"We chose them from the people we felt the most comfortable with," she explains. These were couples in which the woman was usually a homemaker and the man was an accountant, serviceman or salesman—all lower-middle-class, if categorized only by occupation. As friends, these couples hung out together in the recreation hall for birthday, Christmas and Halloween costume parties. The women visited each other every day, shared the pies and cakes they baked, went shopping together and helped each other with housework and baby-sitting.

"There was one woman up the street," my mom adds, "who we were friends with. She associated with us even though she lived in a house."

"Who didn't you feel comfortable with?" I ask her.

"Couples who did a lot of drinking. People who had messy trailers and didn't keep up their yards. People who let their babies run around barefoot in dirty diapers. But there were very few people like that in our trailer park." They were also the boys who swam in the polluted brine of Newark Bay, and the people who trapped crabs in the same waters and actually ate their catch.

When we first entered the social world of Sunset Trailer Park, our family found ways to fit in and even "move up" a little. Physical location was important. Our trailer was first parked in a middle lot. We then moved up by renting the "top space"—as it was called—when it became available. It was closest to the houses and furthest from the bay. Behind it was a vacant housing lot, which belonged to the trailer park owner and separated his park from the houses on the street—a kind of "no-man's-land." The owner gave us permission to take over a piece of this garbage dump and turn it into a garden. My dad fenced it in and my mom planted grass and flowers which we kids weeded. Every year the owner gave out prizes—usually a savings bond—for the best-looking "yards." We won the prize several times.

Yet the privilege of having this extra yard had limits. With other trailer park kids we'd stage plays and performances there for our parents—it was our makeshift, outdoor summer stock theater. But we made so much noise that we drove the woman in the house that overlooked it crazy. At first she just yelled "Shut up!" at us from her second-story window. Then she went directly to the owner, who prohibited the loudest, most unruly kids from playing with us. After a while we learned to keep our own voices down and stopped our shows so we, too, wouldn't be banned from playing in our own yard.

My mom made a little extra money —$25 a month—and gained a bit more social status by working for the owner as the manager of his trailer park. She collected rent checks from each tenant, handled complaints and made change for the milk machine, washers and dryers. "No one ever had trouble paying their rent," Mom tells me, adding more evidence to prove the respectability of our trailer park's residents.

Our family also gained some prestige because my dad "worked in TV" as a pioneer in this exciting new field. Once in a while he got tickets for our neighbors to appear as contestants on the TV game show *Truth or Consequences,* a sadistic spectacle that forced couples to perform humiliating stunts if they couldn't give correct answers to trick questions. Our neighbors joined other trailer park contestants who in the fifties appeared on game shows like *Beat the Clock, Name That Tune, You Bet Your Life* and especially *Queen for a Day.* The whole trailer park was glued to our TV set on nights when our neighbors were on. We rooted for them to win as they became celebrities right before our eyes. Even the put-down "Bayonne jokes" we heard from Groucho Marx, Ralph Kramden and other TV comics acknowledged that our town and its residents had at least some status on the nation's cultural map.

My dad wasn't the only trailer park resident who gained a little prestige from celebrity. There was a musician who played the clarinet in the NBC orchestra. There was a circus performer who claimed he was the only man in the world who could juggle nine balls at once. He put on a special show for us kids in the recreation hall to prove it. There was a former swimming champion who lived in a trailer parked near the water. Although he was recovering from pneumonia, he jumped into Newark Bay to rescue little Jimmy who'd fallen off the seawall and was drowning. He was our hero. And there was an elderly couple who shared a tiny Airstream with 24 Chihuahua dogs—the kind pictured in comic books as small enough to fit in a teacup. This couple probably broke the world's record for the largest number of dogs ever to live in a house trailer. You could hear their dogs yip hysterically whenever you walked by.

Trailer park Chihuahua dog collectors, game show contestants, circus performers—lowlife and weird, perhaps, to outsiders, but to us, these were our heroes and celebrities. What's more, people from all over the United States ended up in our trailer park. "They were well-traveled and wise," my mom explains with pride, "and they shared with us the great wealth of their experience." This may be why they had more tolerance for differences (within our whiteness, at least) than was usual in many white communities during the fifties—more tolerant, my mom adds, than the less-traveled people who lived in the houses.

Migrants and the Community

Toby F. Sonneman

Migrant workers are believed to number around 3 million adults and children, working without health benefits. This essay is taken from a book about the lives and experiences of such workers. The author and her husband are themselves "fruit tramps," enduring low pay, low status, and back-breaking work as they travel to different regions where the fruit is ripe for picking. From the reading, migrants appear to be treated by the public in ways similar to the ways homeless people are treated. What are the similarities and differences between these two groups? How would cultural and structural theorists analyze migrant workers' life situations?

One evening, after the children in camp were asleep for the night, a few people still sat outside in the waning light of dusk, quietly talking. The screen door of one of the trailers nearby opened with a creak, and a woman poked her head out.

"Hank, would you run to the store for some bread? We ain't got enough for tomorrow's lunch."

Hank sighed and got up from his chair reluctantly. "That's about the third time today I've been up at that there Tom's Market," he complained. "I swear, those people are gettin' rich off us fruit tramps."

As he headed for his pickup, the other pickers watched him sympathetically. The little store he was driving to, only a mile and a half away, was surrounded by orchards and used mainly by pickers who didn't want to travel five miles to a larger town.

Walter shook his head. "Yeah, I bought three dollars of groceries over there today, and the man only charged me five dollars!" he quipped.

"Well, you know they jack the prices up just as soon as the fruit pickers hit town," Darlene said. "A gallon of milk is thirty cents higher once the pickers get into town. They just figure we got no choice."

"What gets me is the way some of these locals act toward you, like your money ain't as good as theirs." Travis, a small, energetic man, wrinkled his forehead in disgust. "I went into the grocery one time, and I'd been uptown and got me sixty-five dollars worth of food stamps in my pocket. And I'm about to pay for my groceries, when this old lady says to me, 'You'd better go to the back of the line.' I said, 'Why is that?' and she said, 'Those food stamps. I had to pay for those things.' I said, 'Lady, I had to pay for 'em too, and I got every right to 'em.' Hell, I bet that old lady was drawin' her social security too!

"That makes me so blame mad!" he went on. "We're the ones that's harvesting their food and they think they're better'n us just because we have to use food stamps when there ain't no work."

Because most of the interactions between migrants and the community occur in stores, especially grocery stores, these are the main places where each group forms its opinion of the other. Most of the Okies we talked to had an active distrust of storekeepers. We never discovered if the stores really raised their prices when the pickers came to town, but many of the people we worked with believed they did. Smaller markets, which fruit pickers must often rely on because they are more rural, always had higher prices than stores in town, because of various economic factors. Because pickers frequented these stores in

SOURCE: Toby F. Sonneman, *Fruit Fields in My Blood: Okie Migrants in the West* (Moscow, ID: University of Idaho Press, 1992), pp. 171–175. Reprinted by permission of the publisher.

such great numbers, they often felt that they were being taken advantage of because of their needs.

Fruit pickers are especially sensitive to the attitudes of grocery-store clerks. They often feel negatively judged for the work they do—yet this is the very work that the grocery business profits from.

"Those storekeepers look down on us fruit tramps," Darlene told us. Her strong voice had a bitter edge. "You know how when they give you your change back, whenever you got dirty hands from workin' in the fields, they hold their hand six or eight inches from yours and drop your money down into it."

"Yeah," agreed Thelma, a mild-mannered woman who was married to Travis. "Or they lay the money on the counter, like they'd catch something awful if they touched you."

This particular experience of the clerk not putting the change in the fruit picker's hand is one I heard complaints of often, and I experienced it personally. The implication that fruit pickers are too dirty to touch is especially damaging to a picker's pride.

"It's always been that-a-way," commented Daniel, Darlene's husband, trying to explain the phenomenon. "The people in the towns always seem to get the idea that fruit pickers are dirty. Because they may work ten or fifteen miles off from where they live, and rather than to come home and clean up and then go to the store and buy something for the next day's lunch or supper that night, they just stop on the way home. It saves havin' to go back to town. I guess you could call it conservation of energy, but we seem to have gotten an unclean reputation from doin' this."

Fruit pickers must also use stores more frequently because of their limited storage space and refrigeration. Often there are no sanitary facilities in the field, so their hands as well as their clothes can be dirty from contact with fruit, trees, and ladders. Their hands are torn, covered with hundreds of scratches and cuts from the trees. They are tired, hungry, and dirty by the time they come to the store to buy their food and spend their hard-earned money. If the attitude of the grocery-store clerk is hostile, the experience is not soon forgotten.

"For some reason or other, seems like the communities have a poor outlook on migrants," said Daniel's older brother, Walter. With deep lines etched into his face, Walter looked much older than his forty years. Unlike Travis and Daniel, who wore crisply ironed western shirts with blue jeans, Walter's clothes were rumpled and old, as if he had no time for such frivolities as dress. His demeanor was serious and thoughtful as he continued. "I think local people figure we don't have roots or aren't a part of anything. When in reality, we do have roots, very strong roots, and we are a part of something. We're a part of feeding these people who have this contempt for us."

John Steinbeck noted this same contradiction in *Their Blood Is Strong* in 1938. "Thus, in California we find a curious attitude toward a group that makes our agriculture successful," he wrote. "The migrants are needed, and they are hated. . . . They are never received into a community nor into the life of a community. Wanderers in fact, they are never allowed to feel at home in the communities that demand their services."

Migrants feel that they contribute to the economics of the community not only by their work but by the money they spend locally. "You know, this thing about not spending any money in the community is an absolute lie," said Thelma. "Our cars break down and we get 'em worked on; we buy lots of food and gas and other things too when the job is good and pay for a place to stay lots of times."

The perception that they are taken advantage of by the communities they frequent extends to other services that fruit pickers rely on. "We put five hundred dollars into that truck of ours and you wouldn't believe it," Linda related acridly. "All in this little part that fell off, and this crook in Othello charged us an arm and a leg cuz he knew we couldn't get out of that nasty little town without it. I think I would've left and run away back home if I had a home to run to."

In addition to grocery stores, car-repair places, gas stations, and campgrounds, migrants use state or community services for health care. Because of a disinclination to use such services as well as the difficulty of obtaining regular health care on the migrant run, such services are usually used only for a health crisis, and preventative health care is rarely sought out by migrants. Cultural factors add to the Okies' reluctance to seek health care. A suspicion of medical practitioners; a belief that they can often cure themselves (or a more fatalistic belief that they were meant to be sick); and denial of all but the most major illness, combined with a strong work ethic, all contribute to this behavior.

Notes of a Racial Caste Baby

Bryan K. Fair

As noted in the introduction, an African American woman is almost two and a half times more likely to be poor than a White woman (26.6 percent compared to 10.9 percent). This selection is by Bryan K. Fair, an African American man from Columbus, Ohio, who is now a lawyer, a professor of constitutional law, a university administrator, and an author. This is his personal narrative of growing up as one of 10 children to a single mother in a Black ghetto. Which theory of poverty do you think this author advocates? How does the author appear to feel about the welfare system? What lessons can society learn from his experiences?

RACIAL POVERTY

In Columbus, race is largely a proxy for economic status. Some blacks in the city belong to the middle class and have gone to college and perhaps graduate or professional school. A handful of black doctors and lawyers have practices, and some blacks operate small businesses, on Mount Vernon, Long, Main, or Livingston. Others include teachers and ministers. But most blacks live in one of the ghettos. It is hard to separate race and poverty in black Columbus; they have blended into racial poverty. By contrast, many whites belong to the middle or upper class, living in essentially all-white communities like Bexley, Upper Arlington, Berwick, or Westerville. Thus, if you are black, you aren't expected to be there, especially after dark, and the local police see to it that suspicious (that is, black) people are questioned.

For most of my childhood, my family required Aid for Families with Dependent Children (AFDC), or welfare. But it was not nearly enough for food, housing, and clothing. Part of the welfare money was spent immediately on food stamps. Under the food stamp program, a welfare recipient can buy an allotment of food stamp coupons for less than their face value. For example, one can buy $200 in food stamps for about $125. That may sound like a windfall,

SOURCE: Bryan K. Fair, *Notes of a Racial Caste Baby: Color Blindness and the End of Affirmative Action* (New York: New York University Press, 1997), pp. 14–26. Reprinted by permission of the publisher.

but anyone who has lived on AFDC and food stamps, especially in a large family like mine, knows that the stamps and food always run out well before the end of the month.

For the first two weeks of each month we had milk and cereal, peanut butter and jelly, and plenty of bread. For dinner we ate spaghetti, chili, or Hamburger Helper. Occasionally, we had pot roast or stew. As the month passed, we ate lots of liver and onions over rice, neck bones, or navy beans and corn bread. We also ate luncheon meat, hot dogs, and beans. Except for watermelon on the Fourth of July and oranges at Christmas, we didn't eat fruit. We rarely had balanced meals.

Balanced or not, all too soon the food ran out. Sometimes with ten or more days remaining in the month, we had no more food stamps. The menu then consisted of mayonnaise, sugar, and syrup sandwiches. Among the eleven of us, we could almost always scrape together enough change for bread. We flavored it with whatever we found. Occasionally, we drank sugar water to fill our stomachs. Once, while foraging through the cabinets, I found what I thought was a chocolate candy bar, which I hastily gulped down. But it turned out to be ExLax, a laxative, with predictable, counterproductive results.

If our school had free breakfast or lunch programs, we had at school what was, on some days, our only meals. At other times, I ate at the homes of schoolmates. But sometimes I was too ashamed to accept their parents' offer, even though my stomach ached from hunger. Ironically, few people talk about the stigma or shame of poverty. Instead, they talk about the supposed stigma or harm of remedial affirmative action that can eliminate poverty. Without AFDC, some free meals at school, and generous friends and their parents, I would have gone hungry more often and would have been even more malnourished.

That was true for my siblings as well, but we rarely talked about it. Beyond anger, frustration, and hunger, an empty stomach provides little incentive for conversation.

All across the United States every day, millions of people wake up and go to bed hungry; such is life in "the richest nation in the world." Many days and nights, I sat up hoping that Dee would bring home something, anything, to eat. The voices of teachers talking about the importance of abstract educational skills were often hard to hear over the distracting growling of my stomach.

Even with welfare, there was not enough money for basic clothing or adequate shelter for my family. Dee never had money for clothing or shoes. We didn't have hand-me-downs either. I will never forget the humiliation of either having to wear my sister's shoes to school or not go. Local charities occasionally gave clothing and shoes to needy families. Charity Newsies was one of the largest aid groups in Columbus, and every couple of years, Dee took some of us to a local warehouse where Charity Newsies stocked its clothing: shirts, pants, coats, and shoes. We could pick out three outfits, a coat, and shoes. Usually the selection was limited to two or three styles, colors, and patterns. We reluctantly wore these clothes to school, knowing that some of the kids would recognize where they came from. Nevertheless, we wore those clothes until they were tattered. Even though I was always embarrassed when we went to the warehouse, I was grateful for the charity's help. As I grew older and began working, I proudly bought my own clothes.

After buying food stamps, most of the rest of the welfare check went to pay part of the rent. Rent in the ghetto is artificially inflated because so many blacks are competing for the same slum tenements and because few can afford to live elsewhere. Rich people rarely allow

low-income housing in their communities, using zoning restrictions to keep out undesirable housing and people, a practice that also artificially inflates the value of their property. Zoning is another way in which the law upholds white privilege.

My family lived almost exclusively on the east side of Columbus, in older housing. We almost always lived in houses because apartments were simply too small. Occasionally, we lived in houses owned by the local housing authority. We had to move often in search of cheaper housing, in some cases when we couldn't get any more extensions from the landlord. In my twelve years of school in Columbus, we moved seven times, all within the same ghetto.

Each house that we lived in was at least fifty years old, really just a shell providing only minimal protection from the elements. Some had leaking roofs, others holes that welcomed rats and field mice. None was large enough to accommodate eleven people. We always shared rooms and sometimes beds, depending on the number living at home. I shared a room with my two younger brothers until I moved in with my oldest brother and his wife during my final year of high school.

Our large number was a test for any house, but much greater for one that was small, poorly insulated, and infested with roaches and rats. We had no money for repair, maintenance, or exterminator services. If the landlord didn't pay for them, we did without them. At night, after most of the lights were off, the roaches took over. When we switched on the lights, they scattered for cover. Sometimes, a rat or mouse scurried past in the shadows cast by the blue light of the rented television. When we saw a rat, Dee would bring home traps and place them strategically around the kitchen, some more conspicuous than others. Eventually, while checking the traps, one of us would discover a rat still struggling to escape, or after a few days, the stench confirmed a kill.

Dee never had enough money to pay our high utility bills. She regularly did mental gymnastics to determine which bill was the most urgent, nearest the disconnection stage. Telephone service was a luxury we could only occasionally afford, and our service was often restricted to local calls. After it was disconnected, the telephone company required a deposit and reconnection fee. To avoid these costs, Dee arranged for someone to give her a gift telephone package that had no additional fees. Eventually the phone company recognized the scheme, and we went without a telephone.

Money was sometimes so short that we had no gas heat or hot water during frigid Ohio winters. At those times, racial poverty became surreal. How could there be no heat when the temperature was below zero? The only way to stay warm then was to wear layers of clothing, even in bed. I dreaded the mornings. Each time I stepped into the icy shower, I was sure I would break my neck from jumping around. In the summer a cold shower is bracing but refreshing; in the winter it leaves you trembling, almost numb. I often brooded over how little we had.

Racial poverty was the constant feature in my life in black Columbus. At an early age I was an expert on racial poverty. When I should have been learning to read, write, and play with other kids, I was hungry, cold, and learning about being black. When I should have been developing cognitive skills, I was learning that if a person is hungry enough, he or she will steal for food. Before I could read, I learned how to case a store, locate its clerk, grab something to eat, and walk out without paying. Jequeta's, M&R's, Demler's, Lawson's, Big Bear, Kroger's, and other stores along Livingston or Main were my training ground. Occasionally, when I slipped up, clerks chased me through streets and alleys. I'm lucky no one ever shot at me. I never wanted to steal; it was not a game. But sometimes I was so hungry that I didn't know what else to do.

At times, we look to others to blame for our difficult circumstances, and I did, too. I was ashamed of my racial poverty. But I didn't blame the "white man," and since I didn't know my father, I blamed Dee. For years my relationship with my mother was strained because I blamed her for not telling me who my father was and for all that we didn't have. But Dee was not responsible for my father, and she wasn't lazy or idle. She had graduated from high school and had gone to work. But not every occupation was open to her. I didn't understand then that in Columbus and elsewhere black women could work in only certain jobs.

Undoubtedly, some readers will ask why Dee had ten children. I don't know why she did. But I do know that some states have restricted access to contraceptives and have criminalized abortion. And I do know that Dee's family did not have access to general health care. I assume she couldn't afford or didn't want to have an illegal, back-alley abortion. I also suspect she didn't realize she would have sole responsibility for all of us. As her eighth child, I'm glad, of course, that she didn't stop at seven. Yet even if I believed that having ten children was a bad decision—a point that I'm not willing to concede—it was a decision aggravated by her educational and employment limitations: Dee did women's work for women's wages, and any number of jobs for which she might have qualified were simply closed to her. Dee was not to blame for those constraints. The costs of gender tracking and the devaluation of women's labor are enormous for families headed by single women, forcing most of them into a standard of living below the poverty line. For minority women, race and gender discrimination intersect, pushing even more families into poverty.

Those of us living in the ghetto were not destined for poverty because of inferior genetics or poor values. Rather, we were poor because we had limited opportunities as blacks and because of our circumstances. When I left the ghetto in the seventh grade, my life took a new direction. Escaping black Columbus also taught me that whites are not superior to blacks in regard to genetics or values. If you locked whites in ghetto conditions and denied them rewarding opportunities, most of them would be poor, too. Indeed, in pockets of the United States, some of the poorest citizens are white. It is the self-serving racial rhetoric of the white elites in America that has kept blacks and poor whites, including many white women, divided, largely by pitting them against each other and convincing poor whites that by virtue of the color of their skin, they too are superior to blacks.

Critics often charge that welfare itself creates dependency, a peculiar claim because AFDC was started in the 1930s for the opposite reason: to ease the dependency of poor children. To qualify for it, one already had to be dependent. How, then, can welfare create dependency? Some of welfare's critics believe that the current guidelines increase the number of households headed by single women by threatening to terminate benefits if a father or adult male lives in the home. If this is the case, why not change just that part of the policy? Do these critics believe that welfare payments are so large that rational people would rather receive welfare than work? If so, then why don't more Americans choose welfare's "easy street?" Do the critics think that welfare recipients have the luxury of choosing either to work for poverty wages in dead-end jobs with no medical benefits or to receive public assistance with limited benefits? Or do they think welfare is a legitimate option because some Americans don't earn a living wage?

AFDC did not make Dee idle or dependent; she needed it for her dependent children, not herself. Dee did what so many others have done in similar circumstances: she turned to a government that claims not to permit dependent children to go without food, clothing, or shelter. Unfortunately, most of the time we were on welfare, we did not have sufficient food, clothing, and shelter.

Welfare programs currently provide only marginal subsistence for the millions of children they are supposed to aid. The first priority of welfare reform, therefore, must be to ensure that the money spent adequately feeds these children and puts clothes on their backs. In 1990, the federal government spent $14 billion on the Food Stamp Program while spending $10 billion on criminal justice. In 1996, fourteen million people (two-thirds of them children) received AFDC, for which the federal and state governments spend nearly $23 billion a year. The money we invest in our children's health and educational development now is far less than what, without the investment, we would spend for additional law enforcement and prisons.

Despite the prevailing assumptions, federal spending on AFDC amounts to less than 1 percent of the national budget; yet some politicians and the media lead us to believe that once we "fix" welfare, the huge federal deficit will disappear. Unfortunately, many of the people talking about welfare seem to know very little about it and seem not to care about our most vulnerable citizens: poor children. If they really do want to reform welfare, they should talk to the real "experts": those kids going without food at the end of the month.

I now regret feeling ashamed and embarrassed about my family's racial poverty and especially rue blaming my mother for it. Only now do I understand how hard it was for her to see us live as we did. Unfortunately, we are not born with knowledge and skills to negotiate such feelings, nor are there enough accessible institutions to help poor people cope with such feelings or provide them with guidance to move out of racial poverty. Plans for welfare reform must increase the funding and accessibility of such agencies—such an agency could have helped my family a great deal. Under our extraordinary circumstances, Dee was the best mother we could have had. Although she was smart and talented, she had accumulated few material possessions to show for it. Nonetheless, she fashioned an existence for us out of almost nothing, and she never seemed consumed by despair, hers or ours. As a result, I never lost hope. She never talked about hating whites or the men who had abandoned her and her children. Although we had very little, Dee taught us to take care of what we did have, and she made it clear that she expected each of us to help the family whenever we could. I was my siblings' keeper. We all helped out, and Dee stayed and struggled with us. I don't know why she didn't leave, but it never once occurred to me that she might.

4

Parenting in Poverty

Two out of three impoverished adults in the United States are women, a consequence of the prevailing institutional sexism in society. With few exceptions, U.S. society provides poor job and earnings opportunities for women. This fact, combined with the relatively high frequency of divorce (85 percent of children of divorced parents live with their mothers) and the large number of never-married women with children, has resulted in the high probability of women who head families being poor (24.7 percent of female-headed families with no husband present were below the poverty line in 2000, compared with 4.7 percent of married couples living in poverty) (U.S. Bureau of the Census, 2001:2). Women with children with no husband present were, on average, $7,018 *below* the poverty threshold in 2000 (U.S. Bureau of the Census, 2001:13). In short, the highest risk of poverty results from being a woman and having children, a trend that economist Nancy Folbre has called the "pauperization of motherhood" (Folbre, 1985). This generalization is even stronger when race/ethnicity is considered with gender—single African American and Latina women with children fare worse than White women with children.

Mothers without partners are faced with the quadruple burden of earning a wage, paying the bills, doing the housework, and parenting.

> These families are highly susceptible to poverty, because they typically rely on the earnings of a single adult and incur expenses for child care. Because child support enforcement is poor, few receive substantial transfers from

absent parents. Because levels of public assistance are low, many live in poverty even if they are "on welfare." (Heinz, Folbre, and the Center for Popular Economics, 2000:80).

Parenting by the poor is, of course, not limited to women. Poor fathers may also be single parents facing the often conflicting demands of work, survival, and parenting. Fathers in poor two-parent families often work at two jobs to make up for low wages and benefits.

As you will recall, the first element in the sociological imagination (see the overview in Part I) is a willingness to view the social world from the perspective of others. As you read the essays by and about poor single mothers, imagine what it is like to be in that situation with few resources, the responsibility of parenthood, and working at dead-end jobs. What policies could the government institute to give these people hope for a better life? Or, alternatively, does the government have a responsibility to help them?

NOTES AND SUGGESTIONS FOR FURTHER READING

Albelda, Randy, and Chris Tilly (1997). *Glass Ceilings and Bottomless Pits: Women's Work, Women's Poverty*. Boston: South End Press.

Andersen, Margaret L., and Patricia Hill Collins (eds.) (2001). *Race, Class, and Gender: An Anthology*, 4th edition. Belmont, CA: Wadsworth.

Baca Zinn, Maxine, and Bonnie Thornton Dill (eds.) (1994). *Women of Color in U.S. Society*. Philadelphia: Temple University Press.

Berrick, Jill Duerr (1995). *Faces of Poverty: Portraits of Women and Children on Welfare*. New York: Oxford University Press.

Butler, Sandra S. (1994). *Middle-Aged, Female and Homeless: The Stories of a Forgotten Group*. New York: Garland.

Children's Defense Fund (2001). *The State of America's Children: 2001*. Washington, DC.

Choi, Namkee G., and Lidia J. Snyder (1999). *Homeless Families with Children: A Subjective Experience of Homelessness*. New York: Springer.

Connolly, Deborah R. (2000). *Homeless Mothers: Face to Face with Women and Poverty*. Minneapolis, MN: University of Minnesota Press.

Edin, Kathryn, and Laura Lein (1997). *Making Ends Meet: How Single Mothers Survive Welfare and Low Wage Work*. New York: Russell Sage Foundation.

Folbre, Nancy (1985). "The Pauperization of Motherhood: Patriarchy and Social Policy in the United States." *Review of Radical Political Economics* 16 (4).

Gallay, Leslie S., and Constance A. Flanagan (2000). "The Well Being of Children in a Changing Economy: Time for a New Social Contract in America." Pp. 3–33 in *Resilience Across Contexts: Family, Work, Culture, and Community*. Ronald D. Taylor and Margaret C. Wang (eds.). Mahwah, NJ: Lawrence Erlbaum Associates.

Heintz, James, Nancy Folbre, and the Center for Popular Economics (2000). *The Ultimate Field Guide to the U.S. Economy: A Compact and Irreverent Guide to Economic Life in America*. New York: The New Press.

Kozol, Jonathan (1988). *Rachel and Her Children: Homeless Families in America.* New York: Crown.

Schein, Virginia E. (1995). *Working from the Margins: Voices of Mothers in Poverty.* Ithaca, NY: Cornell University Press.

Sherman, Arloc (1997). *Poverty Matters: The Cost of Child Poverty in America.* Washington, DC: Children's Defense Fund.

Sidel, Ruth (1996). *Keeping Women and Children Last: America's War on the Poor.* Baltimore, MD: Penguin.

U.S. Bureau of the Census (2001). "Poverty in the United States: 2000." *Current Population Reports.* P. 60–214. Washington, DC: U.S. Government Printing Office.

The Empty Christmas

Tamicia Rush

This essay was written as a class assignment by an African American single mother at Olive-Harvey College in South Chicago. In her essay, she remembers a painful Christmas when she was just seven years old. As you read this selection, use your sociological imagination to analyze her situation. How would the different theoretical perspectives view the mother's and father's behavior? Think about the line between private family life and public institutions. At what point, if any, should society intervene in private family life?

The night before Christmas, the air was full of anticipation. My seven-year-old face was shining like the sun, awaiting the return of Santa Claus or, as I knew him, my parents to bring any kind of gift.

I sat in front of the tree that I had put up days before. It was not the best tree. It stood about six feet tall, but leaned against the window like a wreath because there was no stand. The tree had many branches missing, so it was very hard to make it look nice. The lights on that tree were very big and bulky because they were outdoor lights, but they were the only lights in the house. In all, there were about ten ornaments on the tree, which were shedding their thread like a snake shedding its skin. The tree was bare of tinsel and garland because my mother was not there to buy any.

My mother was not there most of the time. She would go off on little adventures and when she came back she acted like she had not been anywhere. When she was gone, there would be no food to eat and no parental supervision in the nasty house. This particular time she went to a friend's house where she said my gifts were. I waited from December 22 to Christmas day for her to return.

When I woke up Christmas morning and looked under the tree, to my surprise there was nothing there. I really believed that I was going to get something this time because she had told me that this Christmas was going to be different. My mother was not even in her room. I went and sat on the sofa and looked at the empty space under the tree and felt the emptiness in my heart. I cried for about an hour. My eyes were blood-shot red and my nose was running like faucet water on a winter day. I felt like no one cared.

As I reached over and got a piece of tissue, the door-bell rang. I went to

SOURCE: Unpublished essay by Tamicia Rush. It is published here with her permission.

the door and opened it, thinking that it was my mother ringing because she had lost her keys. But it wasn't, it was my father. He came in, said "Merry Christmas," and gave me a kiss. He asked me if I had been crying, I told him yes, and then he asked "why?" I said because I hadn't eaten anything for three days and I didn't get anything for Christmas. He told me not to worry. I didn't even have to tell him that my mother wasn't there, he just knew. A blank look came over him, as if he was planning something. He turned and left out the door and, when he came back, he had a bag in his hand. He gave me the bag and in it was a pair of boots and a glove set. It wasn't what I was looking for, but it would do.

My dad and I sat and talked for a minute and then my mother walked through the door. I immediately looked down at her hands and they were as empty as a night sky without stars. She did not speak, she just went straight into her bedroom. She was not alone. My father got up, gave me a couple of dollars, and left. If he hadn't left, he would have been fighting.

After her friend left, she didn't say why she was gone for three days or why there weren't any gifts. She went on pretending like nothing happened and so did I. But, in my heart, I will never forget that Christmas.

Karla

Steven VanderStaay

Introduction by Steven VanderStaay:

"We'll start with what happened," Karla begins. Young, bright, a mother of two children, she is part of a growing phenomenon: single-parent mothers on assistance who cannot afford a place to live. Cities handle such families differently. New York City houses many women like Karla in so-called welfare hotels, while other cities place them in barrack-style emergency shelters or housing projects. Karla could find nowhere to go. Karla is an African American in her twenties. We met at a Salvation Army shelter for homeless families.

Think about this selection from the two theoretical perspectives. What options does this young mother have? What are the life chances of her children? What role should society take in helping this struggling family? Compare/contrast this selection with the previous essay by Tamicia Rush.

I was working up until the time I had my second baby. I lived with my mom but we weren't getting along. She took care of the first child but the second, that was too much. Then she felt that once I had the children . . . well, her words were, "Two grown ladies can never manage in the same house." So I got on AFDC and went to stay with my littlest girl's aunt.

Well, three weeks ago now, her landlord called and said the building didn't pass inspection. See, the building was infested with bugs and mice.

I didn't have any money saved 'cause I was spending all the AFDC and food stamps on us. I do have qualifications for a lot of jobs, but they're all $3.35. And

SOURCE: Steven VanderStaay, *Street Lives: An Oral History of Homeless Americans* (Philadelphia: New Society Publishers, 1992), pp. 170–171. Used by permission of the author.

it's not worth getting a job where you have no medical or dental insurance, not if you have kids. It's not worth giving up welfare. I would work at $3.35 if they let me keep Medicaid and the food stamps, but they don't. They'll cut you off.

But AFDC's not enough to live on either. I started looking for another place but all the apartments I could afford were just like the one we were living in. It wasn't worth leaving one condemnation to go to another.

Then my daughter's aunt, she moved in with her sister. There was no way I could afford an apartment on my own, not and eat too—and like I said, they were all as bad as the first place. So me and the kids—I have a 3-year-old and a 9-month-old—we just stayed in the building. They boarded it up but we got in through the back window.

There was this older lady that lived next door. We were friends and if she could have helped me she would have. But she already had her four grown kids, plus their kids, livin' with her in a two-bedroom apartment. There's a lot of that these days.

She gave us blankets, though, and I wrapped us up in them. We'd stay outside all day, do something—go to the library or I'd take 'em to the museum. Something. Nights we'd go back into the apartment, light candles, and sleep.

Then it rained real bad. And it was cold. The electrical was off, the gas was off, we were going by candlelight. Mice and rats came out really bad. I woke up one morning and there was a mouse on my 9-month-old's head . . . we couldn't stay in there.

So we went outside, walked around all day. Night came and we slept in a car I found. We were wet and both my kids caught a cold real bad. I took 'em to the emergency room and we slept at the hospital.

The next night we were in this laundromat . . . it was so awful. I was crying, the kids were still sick. And my oldest, Robert, he asked a lot of questions: "Momma, why did we sleep in the car? Why are we outside? It's raining, Momma, I'm cold. I don't feel good."

I couldn't explain. And we had been out for the last three days, never being able to rest. He hadn't eaten anything that night 'cause I didn't have any more money.

Then a man at the laundromat, he gave me $4 to get Robert something to eat. And I stole my baby a can of milk.

Single Parenting in Poverty

Virginia Schein

This selection is from the 1995 book of voices of mothers in poverty by Virginia Schein. Schein's words and observations are interspersed here with interviews of single mothers. Many of the women interviewed are divorced and just getting by. Studies indicate that the majority of women experience a serious decline in their standard of living post-divorce. What are the broader implications of this decline in regard to child support and parental responsibility after a divorce? What role should society take in helping single mothers?

SOURCE: Virginia Schein, *Working from the Margins: Voices of Mothers in Poverty* (Ithaca, NY: Cornell University Press, 1995), pp. 55–68. Reprinted by permission of the publisher.

These impoverished single mothers are called on to make a family life out of almost nothing. Poverty is the stage on which their dramas unfold. It has very few props. The women struggle to raise their children with minimal resources beyond the basics of shelter and food, and even food supplies sometimes run low. In addition, the ghost of the absent father is present, adding to the negative circumstances the women must deal with. The ghost is in the scars that an abusive or addicted father has left on the children. Or the ghost is the empty space in the children's lives that the women work hard to fill up. Their lives are like stone soup—trying to make a lot out of very little.

The average income for the women in the interview sample is $7,802 per year. Slightly more than half of the women are on full county assistance, which includes a cash grant, food stamps, and a medical card. Most of the others work part-time, but still receive food stamps and some medical benefits. Some women receive help with other expenses, such as heating. Two women are, or are about to be, employed full-time.

Total Support

Julie, divorced, with two young children, has been living on full benefits for three years. *"I receive $396 a month from the state. My husband pays $160 a month child support and I get $50 a month of that as a rebate from the state. That's the same budget I've been on all this time. Sometimes you rob Peter to pay Paul and pray Paul is not broke. It's very tight. There is $280 coming out of there for rent. That doesn't leave much to pay for heat, phone, and everything else.*

"I'm still waiting for subsidized housing. I do get heat assistance, which is like $183 per winter. That doesn't last long because this house is old. It is gas heat. I use $60 a month for December, January, and February and pray that March is not too cold.

"I worry that the bills won't get paid because I'm very prideful. It is some worry—the money. We've got a roof over our head and we do have food in our belly. It's just the money—when you think that we are living on really $4,000 per year. There is no money there.

"The best example I can give you about money is I can't go in a store and buy my kids a pack of gum without worrying about where the money is coming from. Fifty-nine cents isn't a lot of money to most people. To me, that's a lot of money. I can buy something else with that fifty-nine cents that they need."

Jessie, thirty-nine and divorced, with two children at home, has also been on full benefits for three years. *"I get $182.50 every two weeks from the state, plus food stamps and the medical. I couldn't make it without the medical. The money is barely enough for rent and utilities.*

I pay $200 in rent. It's not subsidized. I keep one whole check out and some of the next check for the rent. That doesn't leave much for utility bills, much less when the kids need books for school and stuff or even clothes. I don't get any child support.

"Last winter was rough. We couldn't use wood because the chimney wasn't working right. We get fuel assistance, but they only help you so much—about two times. That got me through to January. I still had the end of January, February, and March. It was cold. We have a kerosene heater and we pretty much close off the house and stay in one room. We all sleep in that room and when we are home we stay in that room."

Partial Help

Jean was on full benefits for a year. She earned her GED during that period and is now working part-time. For the last several months she and her five school-age children have been living on her earnings, although she still receives food stamps and a medical card.

"I'm proud of myself. I'm not getting no forty hours a week down there at the plant, but I'm managing. I'm managing to keep the bills caught up and everything."

Marie, with three children under twelve, works thirty hours a week and receives food stamps and a temporary medical card. *"I gross $499 every two weeks. I net $329, that's $658 per month. I'm on subsidized day care, which is $36 a* *month, and subsidized rent, which is $68 a month. I get some help on electric and heat. There's not much left for everything else like clothes, telephone, gas, shampoo, my loan. But I'm a bargain hunter. I shop at Goodwill and I go to the Rescue Mission."*

When Things Get Really Tough

Many of the women said that by the end of the month they are sometimes out of food or money. They are forced to borrow, rely on local charities, or juggle their bills in order to feed themselves and their children and make ends meet.

Marie: *"Usually by the end of the month, before the food stamps come in, I am out of everything. I go to the Food Pantry and they will give me a box and a bag of groceries. If you use up all of your crisis money for heat, you can go to the Salvation Army and they will help you out with fifty dollars.*

"I don't like to ask people for help. I usually wait until the last minute. Only if I absolutely have to have something, then I will swallow my pride. Last week I had to ask my dad for help. I called and said, 'Would you please help me get my son a pair of sneakers?'"

Jessie: *"You always run out of food stamps by the end of the month. That is when you go to the Food Basket and the Catholic Church. They help a lot, they really do. My mom tries to help, too. It is tough.*

"I've looked for every possible source of help. The Food Basket and the Catholic Church help with food and clothes. Lutheran Social Services will help with heat. Children's Aid will help with clothes and toys at Christmas.

"I'm existing. When I look back, I don't know how we made it through last month. But here we are."

Renata has two teenagers. She has held a variety of full-time jobs, but is currently on full benefits. *"I don't do too bad, except for things like bread or milk. If I run low, I ask my mother for something. She will give it to me and I replace it back to her when I get my food stamps."*

Carla works part-time as a telemarketer to support her three children, ranging from an infant to one in sixth grade.

"I worry about the bills that are coming in that I don't have the money to pay for. The bill collectors call you on the phone. I'll say, 'Okay, I'll try.' But if I don't send it in, they keep calling back. I would like to be able to pay my bills on time but sometimes there just isn't any money."

RAISING CHILDREN IN POVERTY: THE PAIN OF NOT PROVIDING

For almost all of the women, the most painful part of these tight budgets is not being able to provide well for their children. Finding the money for basics is difficult. Going to the movies or a fast-food restaurant usually costs too much money. Yet, rarely, if ever, did a woman mention items, such as a new dress or a trip, that she might want. Her energies are focused on providing the basics of shelter, food, and clothing for her family. If she can manage all of this reasonably well, she feels okay. For her, the pain of poverty is what her children cannot have.

Basic Needs

Even some of the most basic things, such as shoes and school supplies, pose financial dilemmas for these single mothers.

Jessie: *"My daughter was crying the other day. She doesn't usually talk about what is bothering her. But that day she did. She said, 'I'm tired of being poor.'*

"The book fair is this week. They want to buy books and I don't have the money to give them. Little things like that really bother me bad. They really eat at me. All their friends can do it, but they can't. All I can tell them is that I don't have the money."

Julie: *"The hardest part of being poor is to see them want, even if it is just a Slinky, and knowing that you can't provide that Slinky."*

Joy, twenty-four, has two preschool children. *"They need beds now. They have cribs, but they are almost out of them. But I have to wait and lay away and pick prices, when they need them now. That's the hard part of it."*

Marie: *"It's hard to explain money problems to children. When they say, 'I need shoes, I need clothes,' and you got to keep telling them, 'You have to wait, just wait.' I know the shoes he has on are ready to fall off his feet."*

Renata: *"It's hard not to get my kids the different things they need. I try to get it, but sometimes it takes a little longer. So far, they understand that if I can't get it right then, if they wait a while, I'll try to get it."*

LaVerne, who has one child, works part-time. *"It's hard to say 'no' to my daughter all the time. 'No, you can't have this. No, you can't have that.' Then they send home book papers from school. She wants to buy books from the school. I don't have five dollars to send to the school for a book. That's kind of upsetting."*

Doing Without: Just Living Is Hard

Just living is hard in poverty. The women and their children are surrounded by places, like restaurants, and events, like local fairs, to which they can't go because there is no money. Holiday presents are few or sometimes nonexistent.

Amelia, who lives in a small town, describes what she does with her two children. *"The hardest is trying to take my children places they want to go. I can't afford it. It's that simple.*

"There was a fair here three weeks ago, but there wasn't any money for the tickets. I put the window up and we could hear the music and the rides. We walked over there and watched the rides going. I thought, 'Oh, my god, this hurts.'

"They built a new restaurant here on the corner, a fast-food place. But I haven't been able to take them because I don't have the money. I told the kids—the monies will come in, then the rent and the bills go first. Then you guys go second, all the school stuff and supplies you need. Then, whatever is left over is for entertainment. Let me tell you, there is nothing left."

Carla, living in a city, experiences similar difficulties. *"My kids say, 'Mom, we feel like going out to eat tonight' or 'we want to do this or that.' I would like to have the money to say, 'Okay we'll go,' instead of saying, 'Well, we can't this week,' or 'we can't next week.' I just keep putting it off, but there is never any extra money for such things."*

Holidays, such as Christmas and birthdays, were mentioned frequently as especially painful because of the lack of extra money.

Jessie: *"Christmas coming up this year is a scary thing. I told them yesterday—maybe we should just skip Christmas this year. Maybe put it off until June. They are at the age when they know there is no Santa Claus. They know it depends on me. It's real tough when they believe in Santa Claus because they don't care if you are working or not. I would like to believe in Santa Claus again myself."*

Nancy, with three children in elementary school, works part-time as an occupational therapist. *"It's hard with three little kids. They want so much. Any toy they see on TV, they say, 'I want that.' With Christmas coming up you want to try and get them some of the things they want from on TV. It is like all your money goes for Christmas and you don't have nothing left for now."*

LaVerne: *"My daughter's birthday is coming up. I haven't got her a thing yet. I have to pay the dentist one hundred bucks at the end of the month. Her birthday is coming up right after that and I'm not able to get her anything. So I'm going to try to borrow some money somewhere. But you can never get ahead, that's the thing. The worst part about the money situation is you just can't get ahead."*

A Bleak Future for the Children

Carla, twenty-six, is raising three children. *"I think a lot about my oldest and what her future is going to be like. I want her to have an education. High school is nothing these days. You have to have more than that. That worries me. That I have nothing in the bank for her.*

"She wants to be a teacher. She talks about it all the time. She's a little young and that might change by the time she graduates. She definitely wants to go to college. She stays on the honor roll, distinguished honors. She's a real good student. She's in sixth grade now. I really don't think that I am going to be able to get a job that will pay all of my bills, for all of us, plus put money in the bank for her to be able to go to college. I would want all of my three children to go, but especially her because she's the one that's talking about it.

"It's terrible if your child wants an education and you can't do that for them."

As these voices reveal, the children hear "no" a lot. And they hear "no" for things that "all their friends and everyone on TV" seem to have. All parents hear this lament, regardless of economic circumstances. Yet these mothers can't afford some of the smallest things—new shoes, a night at McDonald's, or a Chicago Bulls cap. These are luxuries.

Single parenting has its stresses and strains. There is no one else to share the fears and bear the responsibilities. But these mothers also live with the overarching worry of money. Will the bills get paid? What if we run out of food? What if something happens? What if I lose this job? What if the rent goes up or my car breaks down? There is no extra money to put away for a child's education. The women struggle just to support their children on a day-to-day basis.

RESOURCEFUL CHILD RAISING

Despite their restricted circumstances, many women managed, sometimes very creatively, to find things to do with their children that don't cost much money. Resourceful child raising is what it takes and the women manage to have fun with their children, perhaps putting their money worries on the back burner for a day.

Julie: *"In the winter we have picnics in the living room. We put the sheet on the floor. We make hamburgers. We make potato salad. We have raisins—the whole nine yards. We sit and have a picnic.*

"You learn to do things free. You find out what is free. If there is a reading program that's not too far away and there's enough gas in the car, you go."

Emily: *"On Monday nights we have Brownies for Betty. We have Girl Scouts for Kim on Tuesdays. Wednesday we go to church. I try to do something everyday to get us out of the house. We do the library all the time. Library programs are free. I've searched out all the different free things there are to do."*

Vicki: *"The carnival and the fair are coming up—we will probably go to that. Bowling, miniature golfing, we do that. There is not a whole lot we can do because of the money. We just spend most of our time here at home together, doing homework, reading, or watching TV."*

Amelia: *"My son plays baseball, and we go to all the games here. I've got a Brownie troop this year. If we can, we go to the lake, have a picnic, and stay through lunch and dinner, so they can swim all day. We have a cat and a dog. We wouldn't give them up, either."*

Arlene: *"I try not to sit around and worry about the money because life's too short. I try to have fun. I will take a little bill money every now and then and take the kids to the arcade or to the seventy-five cents movie or just spend a day at the mall. We try to do something and have a little fun because my children and I are not going to miss out on life just because I am in the situation I'm in."*

Renata: *"I had overtime pay for a whole week straight. When I came home that Friday my paycheck was bigger than it had ever been. Me and my kids, we went out. I took them to dinner at Pizza Hut. My son wanted a little pan pizza just for himself, and I bought him that. We had a good time. We caught the bus up and laughed and played on the bus and we caught the bus back. They enjoyed it."*

When Mommy Gets A (Better) Job

Several of the women were in school or in training programs that gave them hope for a better financial future. Their children's reactions to their schooling or jobs were often expressed in terms of having more money.

Connie, who is pursuing a degree in nursing, has two children, ages seven and nine. *"When we go to the store they say, 'Mom, can we have this one day when you get a whole bunch of money?' Or they will say, 'When you are through college . . .' That's what their famous line is because I always tell them, 'When I'm through with college . . .'"*

Susan's five-year-old son is a supporter of her efforts. She is newly certified as a nurse's aide. *"My son is always saying, 'You have to go back to school or you have to do that.' Every once in a while he sees a red sports car and he'll say, 'When you get to be a nurse, you can have that.'"*

Gloria received her associate college degree recently. *"Sometimes my kids say, 'Mom, why can't you buy me this?' or 'Why can't you buy me that?' Now I say, 'Just wait, some day.'"*

These mothers are not different from other parents. They, too, want happy Christmas and holiday mornings and birthdays for their children to remember. Their creativeness and efforts to find fun things to do are ways to make the stone soup tasty.

DEALING WITH DADDY

The fathers are rarely present in these families. The absence of a father who was abusive or who suffered from alcohol or drug addictions is a help. But the women have been left to deal with the aftermath. The men are gone, but their ghosts are in the psychological scars left on the children.

Even if the children did not experience traumatic incidents, the women must explain the void. "Why did daddy leave?" "Where is daddy?" "Doesn't he love me?" are questions the women handle all the time. To be a family, the women have to explain his absence and abuse in ways that minimize the pain of his loss. "Dealing with daddy" is added to the already difficult work of raising children alone and in poverty.

Healing the Wounds of Abuse

Amelia's two young children watched as their father beat her up. She describes the effects of this abuse on her children: *"My older child seems okay. 'I've got my mom, that's all I need.' But the abuse hit my younger child much harder. She clams up in this little shell and she won't come out.*

"She drew these pictures of her family life for the counselor she's seeing. Myself and her brother are on one side, on the swings and walking the dog. On the other side she drew her father, in a big, big bucket. 'I don't have a daddy,' she said.

"Yesterday she came home and she did four hours of homework to catch up in school. She will do her work, but she won't pass it in. I asked her why and she said, 'I'm afraid it is not good enough.' Sometimes I don't know what to do. I just hate to see her hurting like that. But the school has been very understanding and the counselor is very helpful."

Gloria's husband abused her and their five children. He raped and molested her daughters. She helped her children heal their personal pain. *"My kids are great. My oldest daughter went through some rough times, but now she is married and has a little girl. My son also went through some pretty rough times. He had a lot of anger from a lot of the things his father put him through and put us through. He's back in school and is thinking about college.*

"My three younger kids have adapted pretty well. Most of the problems they have had has been with their stomach—irritable

bowel syndrome and ulcers. It's from all the stress they went through. They all worked with counselors. They were doing poorly in school, but they have brought their grades up. They are much happier kids now.

"They don't talk about their father at all. Two of my children have taken my maiden name, as I did. The three that are home want to change their name. It costs like $250 for each of them to do it and I don't have it. I'm going to try to get it done, one each year before they graduate so they can graduate with the 'R' name."

Sally has experienced two tumultuous and abusive marriages. *"My husband walked out when my daughter was five. He had a terrible temper. He always made it a point of fighting in front of my daughter when she was a baby. He wouldn't fight any place else. Finally I told him to leave.*

"The second time I was married a year. I asked him to leave, too. He was abusive a couple of times to me. But what really broke the straw was when I found out his son was abusing my daughter. I might have taken it for a while, but my kid, no. To this day she don't want to talk about it.

"She has a temper that is hard to control. Once we were arguing. She wanted to leave home. I tried to stop her. She came after me with a butcher knife. She ended up in the hospital and the whole family went into therapy. She is doing better now in school and with her temper control."

The children of Amelia, Gloria, and Sally were all involved in very traumatic situations. Amelia, Gloria, and Sally deal with their children's behavioral and physiological symptoms while relying on a patchwork of counseling help. At the same time, the women have their own pain to heal as well.

As Amelia stated: *"I've gone through it and I'm trying to help her. Yet I know myself that I am still trying to recover from it. I'm still hurting inside. But what I want most of all is for her hurting to stop. I want her to grow up as a normal little child. But Mr. Smalley (her counselor) said it is going to take years and years."*

Filling the Void, the Best Way You Know How

Even without severe trauma, the father's absence and abandonment leaves a void that the mother must try to cope with and explain.

Carol was in an eight-year relationship with her daughter's father. He walked away from them a few years ago, when the girl was four years old. Carol describes her little girl's reaction today. *"She clings to me like you would not believe. She has this deep-seated fear of abandonment. She has even seen a psychiatrist about it. When I do anything on my own she asks, 'Mom, are you trying to get rid of me? Mom, don't you like me?' Trying to convince her all the time is kind of hard."*

Most of the mothers worry about how to explain daddy's absence to their children. They worry not just about now, but about how their children will react when they are older.

The father of Anita's little girl hasn't seen her since she was two months old and has no interest in the child. *"I don't want her to feel demeaned by him in any way. I try not to say anything. Now one day when she is older she is going to ask you know questions. She has already asked, 'Where does he live?' I told her, 'Millerton.' She says, 'Is that far away?' I say, 'No, not really.' She says, 'Well, why doesn't he live with us? Did you get into a fight?' It gets really hard to answer these questions.*

"I don't want to bring it up. If she brings it up, fine. One day she will ask where he lives or whatever. If I know, then I'll tell her. It's her decision, not my decision. If she wants to be part of his life and he changes his mind, who knows? Maybe he will be sorry for all that he missed. But I'm not missing out on anything, he is. So if one day she wants to know where he is and I know, then she can go and see him. She can see for herself. She has to make up her own mind."

Connie's ex-husband sees his two children very rarely. *"Their father came and took them to the movies twice and they still bring that up. 'What about dad? Can't he take us here or there?' I try to explain why their father doesn't come. I try to be honest. 'Daddy is angry. When daddy gets over being angry, he will come around.' They deal with that. It seems better then lying to them, saying he is away or working."*

Susan struggles with the effects of the absent father of her little boy. *"I have such mixed feelings about raising him alone. I'm glad that his father and I are divorced. If we would have stayed together, my son would have grew up seeing what his father did. He would have grew up thinking that drinking was cool, that running around was the way to live.*

"Now since we live here and my fiancé and I work, he sees what life is about. I anticipate problems with him when he gets to be a teenager because right now he is very confused. His father doesn't come and get him, doesn't see him unless I take him over there. Now he says it's okay that his dad doesn't want to see him, but he is only five. When he's older he's going to have a hard time understanding why.

"So I don't know. I kind of believe in like the Cleaver family. The mom and dad who are so supportive. That's what I wanted to believe when I was younger. When I first had my son I thought, okay, I'll stay home and be the supermom and dad could go out and work. But that's not the way it was. It can't be like that for me anymore.

"I just hope he grows up on the right track. He was too young to remember when we left his dad. Maybe it won't have as much of an impact as I think it will."

LaVerne, never married, is also trying to do the best she can in explaining the absent father and his broken promises to her daughter. *"The hardest part about raising my daughter is when she asks about her dad. That's real hard. When she was little, he never made any contact. He never wrote a letter. He's been in and out of jail.*

"She really didn't know she had a dad until his mother came up here to visit her. My daughter didn't even know who she was because they don't send her cards for birthdays or Christmas or nothing like that. His mother said, 'I'm your grandma.' She said, 'No, my grandma lives in Baylerville.' She said, 'I'm your dad's mother' and my daughter said, 'No, I don't have a dad.' I never told her she didn't have a dad. He had just never made contact with her.

"He called her up before school started and said he was going to buy her some clothes. Now she has been calling my mom, asking if her clothes came yet. He told her he was going to send her twenty dollars a month allowance. She hasn't seen it and she has been looking forward to it. So there he is putting things in her head that I got to take back out.

"I try to explain it to her. I sit down and tell her all the problems me and him had and his mental problems. Why I don't want her to see him. Why she is better off where she is at. It might be better, but it's the best I can do. But at nine, I don't think she understands.

"I told her that when you get to be fifteen or sixteen and want to visit your dad, that's fine. But he will either come here for an hour or you can see him at your grandparents' home for an hour. But you will never leave with him until you are eighteen. My folks say I'm wrong, but they don't know the whole story. They don't know about the drugs and the abuse."

There are no prescribed ways to deal with these situations. The mothers struggle with just how honest to be, especially when abuse or drug or alcohol addiction was involved. They know, as does Susan, that their children are probably better off in a household not disrupted by alcoholic incidents. But they still feel that tug of "but a child needs a father." Perhaps many think, "Yes that's true, but not this one."

They worry that the abandonment will hurt the child's sense of self. But they also worry about protecting their children from a father who may not be responsible enough to be with his children. The residual of the father takes many forms, none of them easy to deal with. To the woman, it can feel like the tail of a dinosaur swatting her just when she thinks she's gotten away.

Some women, themselves raised in a society that promotes the traditional family, are uncomfortable and pained by the loss of what might have been. Eva has two young children and Arlene has three children, the two younger ones of elementary school age.

Eva: *"It's hard, really hard, especially around Christmastime. It really gets to me then that their father is not around to spend time with them. To play like Santa Claus at the door and stuff like that. 'Hurry up and get into bed' . . . like my grandmother used to do to me."*

Arlene: *"It's hard for me being brave about the complete family. It does make a difference. It comes up in school. They are not like the other little kids that have their mom and dad in the household."*

WOMEN OF HARDSHIP

The struggle with financial issues and the worries and insecurities that accompany that struggle are one part of what single parenting in poverty means. Experiencing the pain of not providing for their children is another important aspect. And dealing with the aftereffects of the absent daddy is also in the pot. It is truly a stone soup. Yet women grapple with all of these and still manage to be a family, and raise their children.

The poverty that these women live with every day can crush the spirit. But most live with a sense of hope that things will get better, a sense of hope inspired by their children. All of the women want their children to grow up and live in better circumstances than they are in now. Some hope that their own positive efforts will be a model for their children.

Not for My Child

Gloria: *"The oldest at home, she wants to go to college. The other one wants to go in the service. My youngest want to work in radiology. A few years ago all they wanted was to get out of the house. That's all they thought about. But now they think they will live at home and work.*

"I think it was me going back to school because I've noticed they have tried to bring their grades up, too. I tell them, 'If I can do it, you can do it.' Plus I tell them that if they want a nice car or a nice home, you have to have money to do it. You can't make that kind of money working in a restaurant or as a laborer. You have to have an education. They see the job market because I talk to them very honestly about it. They watch what I'm going through. They now understand what life is about."

Renata: *"Working makes me feel important and it makes my kids look up to me. 'My mommy is doing this. She is not at home, doing nothing.' I always tell my daughter, 'If you ever leave me, make sure you have your own money, that you have your own job. You don't have to take no stuff from no man because you have your own.' I want her to see that in me."*

Jessie: *"I'm going to keep my girls in school. In school they have career days and people come and show them what they do. Any kind of talent they have, I want them to develop so they have a choice. I'm definitely letting them know that life is not a fairy tale. You don't get married and stay home, and live happily ever after."*

Pat: *"I want so badly for them to not make the mistakes I did. I want them to go to college. I want them to be productive in their life. I want them to be happy over all, no matter what happens. I guess that's what every parent wants for their children."*

The pain of not providing for their children is close to the surface for most of the women. Yet, at the same time, the effort to provide and be a family is what keeps them going. If it means going to the Food Basket or leaning on local charities for food and clothes, then they do it. These women worry about the price of gum, say "no" to a child's wish for a movie or a new dress, feel miserable when they can't buy a school book, and try to have fun doing free things with their children. They juggle exhausting schedules of work, school, and child rearing so that they can get better jobs and improve their lives. They deal with the scars on their children left by the dad and explain his absence in the best way that they can. There is little time or opportunity to heal their own pain and loss. They work to overcome hardships, forging a life within circumscribed situations.

The Tomalsons and the Riveras: Families on the Fault Line

Lillian Rubin

This excerpt is from Lillian Rubin's book about America's working class in 1994. The Tomalsons are an African American family in New York City; the Riveras are a Latino family in Chicago. Their words are interspersed with the words and impressions of the author. As you read this selection consider the following: What problems does each family confront as a result of its poverty? How has race had an impact on both families? Is having an intact, two-parent family enough to raise healthy, productive children? If not, what other factors are involved?

THE TOMALSONS

When I last met the Tomalsons, Gwen was working as a clerk in the office of a large Manhattan company and was also a student at a local college where she was studying nursing. George Tomalson, who had worked for three years in a furniture factory, where he laminated plastic to wooden frames, had been thrown out of a job when the company went bankrupt. He seemed a gentle man then, unhappy over the turn his life had taken but still wanting to believe that it would come out all right.

Now, as he sits before me in the still nearly bare apartment, George is angry. "If you're a black man in this country, you don't have a chance, that's all, not a chance. It's like no matter how hard you try, you're nothing but trash. I've been looking for work for over two years now, and there's nothing. White people are complaining all the time that black folks are getting a break. Yeah, well, I don't know who those people are, because it's not me or anybody else I know. People see a black man coming, they run the other way, that's what I know."

"You haven't found any work at all for two years?" I ask.

"Some temporary jobs, a few weeks sometimes, a couple of months once, mostly doing shit work for peanuts. Nothing I could count on."

"If you could do any kind of work you want, what would you do?"

He smiles, "That's easy; I'd be a carpenter. I'm good with my hands, and I know a lot about it," he says, holding his hands out, palms up, and looking at them proudly. But his mood shifts quickly; the smile disappears; his voice turns harsh. "But that's not going to happen. I tried to get into the union, but there's no room there for a black guy. And in this city, without being in the union, you don't have a chance at a construction job. They've got it all locked up, and they're making sure they keep it for themselves."

When I talk with Gwen later, she worries about the intensity of her husband's resentment. "It's not like George;

SOURCE: Lillian Rubin, *Families on the Fault Line: America's Working Class Speaks About the Family, the Economy, Race, and Ethnicity* (New York: HarperCollins, 1994), pp. 225–229, 231–233.

he's always been a real even guy. But he's moody now, and he's so angry, I sometimes wonder what he might do. This place is a hell hole," she says, referring to the housing project they live in. "It's getting worse all the time; kids with guns, all the drugs, grown men out of work all around. I'll bet there's hardly a man in this whole place who's got a job, leave alone a good one."

"Just what is it you worry about?"

She hesitates, clearly wondering whether to speak, how much to tell me about her fears, then says with a shrug, "I don't know, everything, I guess. There's so much crime and drugs and stuff out there. You can't help wondering whether he'll get tempted." She stops herself, looks at me intently, and says, "Look, don't get me wrong; I know it's crazy to think like that. He's not that kind of person. But when you live in times like these, you can't help worrying about everything.

"We both worry a lot about the kids at school. Every time I hear about another kid shot while they're at school, I get like a raving lunatic. What's going on in this world that kids are killing kids? Doesn't anybody care that so many black kids are dying like that? It's like a black child's life doesn't count for anything. How do they expect our kids to grow up to be good citizens when nobody cares about them?

"It's one of the things that drives George crazy, worrying about the kids. There's no way you can keep them safe around here. Sometimes I wonder why we send them to school. They're not getting much of an education there. Michelle just started, but Julia's in the fifth grade, and believe me, she's not learning much.

"We sit over her every night to make sure she does her homework and gets it right. But what good is it if the people at school aren't doing their job. Most of the teachers there don't give a damn. They just want the paycheck and the hell with the kids. Everybody knows it's not like that in the white schools; white people wouldn't stand for it.

"I keep thinking we've got to get out of here for the sake of the kids. I'd love to move someplace, anyplace out of the city where the schools aren't such a cesspool. But," she says dejectedly, "we'll never get out if George can't find a decent job. I'm just beginning my nursing career, and I know I've got a future now. But still, no matter what I do or how long I work at it, I can't make enough for that by myself."

George, too, has dreams of moving away, somewhere far from the city streets, away from the grime and the crime. "Look at this place," he says, his sweeping gesture taking in the whole landscape. "Is this any place to raise kids? Do you know what my little girls see every day they walk out the door? Filth, drugs, guys hanging on the corner waiting for trouble.

"If I could get any kind of a decent job, anything, we'd be out of here, far away, someplace outside the city where the kids could breathe clean and see a different life. It's so bad here, I take them over to my mother's a lot after school; it's a better neighborhood. Then we stay over there and eat sometimes. Mom likes it; she's lonely, and it helps us out. Not that she's got that much, but there's a little pension my father left."

"What about Gwen's family? Do they help out, too?"

"Her mother doesn't have anything to help with since her father died. He's long gone; he was killed by the cops when Gwen was a teenager," he says as calmly as if reporting the time of day.

"Killed by the cops!" The words leap out at me and jangle my brain. But why do they startle me so? Surely with all the discussion of police violence in the black community in recent years, I can't be surprised to hear that a black man was "killed by the cops."

It's the calmness with which the news is relayed that gets to me. And it's the realization once again of the distance between the lives and experiences of blacks and others, even poor others. Not one white person in this study reported a violent death in the family. Nor did any of the Latino and Asian families, although the Latinos spoke of a difficult and often antagonistic relationship with Anglo authorities, especially the police. But four black families (13 percent) told of relatives who had been murdered, one of the families with two victims—a teenage son and a 22-year-old daughter, both killed in violent street crimes.

But I'm also struck by the fact that Gwen never told me how her father died. True, I didn't ask. But I wonder now why she didn't offer the information. "Gwen didn't tell me," I say, as if trying to explain my surprise.

"She doesn't like to talk about it. Would you?" he replies somewhat curtly.

It's a moment or two before I can collect myself to speak again. Then I comment, "You talk about all this so calmly."

He leans forward, looks directly at me, and shakes his head. When he finally speaks, his voice is tight with the effort to control his rage. "What do you want? Should I rant and rave? You want me to say I want to go out and kill those mothers? Well, yeah, I do. They killed a good man just because he was black. He wasn't a criminal; he was a hard-working guy who just happened to be in the wrong place when the cops were looking for someone to shoot," he says, then sits back and stares stonily at the wall in front of him.

We both sit locked in silence until finally I break it. "How did it happen?"

He rouses himself at the sound of my voice. "They were after some dude who robbed a liquor store, and when they saw Gwen's dad, they didn't ask questions; they shot. The bastards. Then they said it was self-defense, that they saw a gun in his hand. That man never held a gun in his life, and nobody ever found one either. But nothing happens to them; it's no big deal, just another dead nigger," he concludes, his eyes blazing.

It's quiet again for a few moments, then, with a sardonic half smile, he says, "What would a nice, white middle-class lady like you know about any of that? You got all those degrees, writing books and all that. How are you going to write about people like us?"

"I was poor like you once, very poor," I say somewhat defensively.

He looks surprised, then retorts, "Poor and white; it's a big difference."

THE RIVERAS

Once again Ana Rivera and I sit at the table in her bright and cheerful kitchen. She's sipping coffee; I'm drinking some bubbly water while we make small talk and get reacquainted. After a while, we begin to talk about the years since we last met. "I'm a grandmother now," she says, her face wreathed in a smile. "My daughter Karen got married and had a baby, and he's the sweetest little boy, smart, too. He's only two and a half, but you should hear him. He sounds like five."

"When I talked to her the last time I was here, Karen was planning to go to college. What happened?" I ask.

She flushes uncomfortably. "She got pregnant, so she had to get married. I was heartbroken at first. She was only 19, and I wanted her to get an education so bad. It was awful; she had been working for a whole year to save money for college, then she got pregnant and couldn't go."

"You say she had to get married. Did she ever consider an abortion?"

"I don't know; we never talked about it. We're Catholic," she says by way of explanation. "I mean, I don't believe in abortion." She hesitates, seeming uncertain about what more she wants to say, then adds, "I have to admit, at a time like that, you have to ask yourself what you really believe. I don't think anybody's got the right to take a child's life. But when I thought about what having that baby would do to Karen's life, I couldn't help thinking, *What if . . . ?*" She stops, unable to bring herself to finish the sentence.

"Did you ever say that to Karen?"

"No, I would *never* do that. I didn't even tell my husband I thought such things. But, you know," she adds, her voice dropping to nearly a whisper, "if she had done it, I don't think I would have said a word."

"What about the rest of the kids?"

"Paul's going to be 19 soon; he's a problem," she sighs. "I mean, he's got a good head, but he won't use it. I don't know what's the matter with kids these days; it's like they want everything but they're not willing to work for anything. He hardly finished high school, so you can't talk to him about going to college. But what's he going to do? These days if you don't have a good education, you don't have a chance. No matter what we say, he doesn't listen, just goes on his smart-alecky way, hanging around the neighborhood with a bunch of no-good kids looking for trouble.

"Rick's so mad, he wants to throw him out of the house. But I say no, we can't do that because then what'll become of him? So we fight about that a lot, and I don't know what's going to happen."

"Does Paul work at all?"

"Sometimes, but mostly not. I'm afraid to think about where he gets money from. His father won't give him a dime. He borrows from me sometimes, but I don't have much to give him. And anyway, Rick would kill me if he knew."

I remember Paul as a gangly, shy 16-year-old, no macho posturing, none of the rage that shook his older brother, not a boy I would have thought would be heading for trouble. But then, Karen, too, had seemed so determined to grasp at a life that was different from the one her parents were living. What happens to these kids?

When I talk with Rick about these years, he, too, asks in bewilderment: What happened? "I don't know; we tried so hard to give the kids everything they needed. I mean, sure, we're not rich, and there's a lot of things we couldn't give them. But we were always here for them; we listened; we talked. What happened? First my daughter gets pregnant and has to get married; now my son is becoming a bum."

"Roberto—that's what we have to call him now," explains Rick, "he says it's what happens when people don't feel they've got respect. He says we'll keep losing our kids until they really believe they really have an equal chance. I don't know; I knew I had to *make* the Anglos respect me, and I had to make my chance. Why don't my kids see it like that?" he asks wearily, his shoulders seeming to sag lower with each sentence he speaks.

"I guess it's really different today, isn't it?" he sighs. "When I was coming up, you could still make your chance. I mean, I only went to high school, but I got a job and worked myself up. You can't do that anymore. Now you need to have some kind of special skills just to get a job that pays more than the minimum wage.

"And the schools, they don't teach kids anything anymore. I went to the same public schools my kids went to, but what a difference. It's like nobody cares anymore."

"How is Roberto doing?" I ask, remembering the hostile eighteen-year-old I interviewed several years earlier.

"He's still mad; he's always talking about injustice and things like that. But he's different than Paul. Roberto always had some goals. I used to worry about him because he's so angry all the time. But I see now that his anger helps him. He wants to fight for his people, to make things better for everybody. Paul, he's like the wind; nothing matters to him.

"Right now, Roberto has a job as an electrician's helper, learning the trade. He's been working there for a couple of years; he's pretty good at it. But I think—I hope—he's going to go to college. He heard that they're trying to get Chicano students to go to the university, so he applied. If he gets some aid, I think he'll go," Rick says, his face radiant at the thought that at least one of his children will fulfill his dream. "Ana and me, we tell him even if he doesn't get aid, he should go. We can't do a lot because we have to help Ana's parents and that takes a big hunk every month. But we'll help him, and he could work to make up the rest. I know it's hard to work and go to school, but people do it all the time, and he's smart; he could do it."

His gaze turns inward; then, as if talking to himself, he says, "I never thought I'd say this but I think Roberto's right. We've got something to learn from some of these kids. I told that to Roberto just the other day. He says Ana and me have been trying to pretend we're one of them all of our lives. I told him, 'I think you're right.' I kept thinking if I did everything right, I wouldn't be a 'greaser.' But after all these years, I'm still a 'greaser' in their eyes. It took my son to make me see it. Now I know. If I weren't I'd be head of the shipping department by now, not just one of the supervisors, and maybe Paul wouldn't be wasting his life on the corner."

REFLECTION QUESTIONS FOR PART III

After reading the selections on living on the economic margins, reflect on the following questions.

1. Part III is titled "Living on the Economic Margins." Compare and contrast the readings in this section. Are the different individuals and families living on the economic margins affected in the same way?
2. How do stigma and shame fit into each selection? What survival strategies do the individuals use to cope with their situations?
3. How do gender/race/ethnicity/age/location structure their experiences?
4. Analyze each article carefully and group them by individual/cultural and structural theories of poverty. What theory is best reflected in each reading? Explain.
5. Select one particular essay, such as Susan Sheehan's "Ain't No Middle Class," and analyze it first *without* using a sociological imagination, and then analyze it again *with* a sociological imagination.

PART IV

The Impact of Societal Institutions on Individual Lives

Focusing on the poor and ignoring the system of power, privilege, and profit which makes them poor, is a little like blaming the corpse for murder.

MICHAEL PARENTI (1978)

As we have seen, one view of the poor is that they are to blame for their economic deprivation. From this perspective, the system is good but flawed people fail. This emphasis ignores the social organization that oppresses the poor.

The alternative view is that poverty is not fundamentally a matter of individual behavior but rather a matter of public institutions and accepted practices that create or perpetuate poverty. Sociologists' focus on society's institutions and on society's role in producing and perpetuating poverty is the subject of Part IV. The chapters in this part examine, in turn, housing/neighborhoods, the welfare system, the health-care system, schools and schooling, and work and working.

5

Housing/Homeless Shelters/Neighborhoods

THE LACK OF AFFORDABLE HOUSING

Affordable housing for the poor is in ever shorter supply. **Gentrification** (the redevelopment of poor and working-class urban neighborhoods into upscale enclaves) and other forms of urban development (building highways, stadiums, parks, parking garages) have significantly reduced the number of affordable rental units. As a result the cost of renting has been rising at about twice the rate of inflation (Children's Defense Fund, 2001:x). The government assumes that the poor should not pay more than 30 percent of their income for housing, yet according to the Department of the Census Bureau, 6.4 million households spend more than half their income on housing (Pollard and Mather, 2001). For poor families to keep rental costs under 30 percent of their earnings, they would have to earn nearly twice the federal minimum wage for a 40-hour week to pay the national median fair-market rate for a two bedroom unit (McCoy, 2000).

Less than 30 percent of low-income renters receive any housing subsidy (Children's Defense Fund, 2001:x), leaving only about 1.3 million poor families receiving vouchers—a government rent subsidy that pays the difference between 30 percent of their income and market rents anywhere they choose to live. Another 660,000 are waiting for vouchers, with waiting times estimated to be ten years in Los Angeles and Newark, seven years in Houston, and five years in Memphis and Chicago (*The Progressive,* 2000). Public housing policy (vouchers and building low-cost housing units) has resulted in 4.3 million affordable units, less than half the estimated need (Maggi, 2000). The government's housing policies to aid the poor, by the way, cost about $30 billion a year, while at the same time another government housing policy allows homeowners to save $115 billion a year by allowing interest and taxes on houses to be tax deductible (*The Progressive,* 2000).

The shortage of affordable housing leaves the poor with three unappealing options—paying excessively high rents, living in substandard housing, or homelessness. Sometimes the choice is a forced one because the poor are only an economic disaster (brought on by theft, fire, illness, or job loss) away from being evicted, resulting in homelessness. Currently, an estimated 2 million Americans are homeless at some time during the year, with about 700,000 homeless on any given night (the population of a city the size of Seattle). In 29 major cities, the homeless outnumber the number of beds in temporary shelters, forcing many to spend the night in cars, in doorways, under bridges, or on the streets. More than one-third of the homeless are families with children. Most U.S. cities are unfriendly to the homeless, with official policies banning loitering, trespassing, panhandling, and carrying open containers of alcohol. A number of cities conduct nightly "police sweeps" of their streets to harass the homeless. New York City requires the homeless to work 20 hours a week in exchange for shelter. Sacramento, California, pays for one-way bus tickets out of state for homeless people with families or jobs to go to. Chicago has privatized sidewalks in front of businesses, which means that anyone who loiters there is trespassing. Thus, the homeless are twice cursed—they are not only without shelter, they are also punished for not having a shelter.

LIVING IN POOR NEIGHBORHOODS

The poor cluster in neighborhoods where the housing is cheap. These places are cheap because of their undesirability. Often these areas are exposed to environmental hazards such as toxic wastes, high concentrations of lead, and other pollutants (Bullard, 2000). They may be cheap because they are in a flood plain or near a noisy highway. Poor urban neighborhoods, typically, are not provided the same services found in more affluent neighborhoods such as banks, retail outlets, and adequate public transportation. Neighborhood schools are often in disrepair and are not equipped with adequate libraries, computers, and special equipment for children who are disabled.

The city neighborhoods where poverty is concentrated have relatively high rates of street crime, making the neighborhoods relatively unsafe. When good job opportunities are scarce, young people in the inner city become more susceptible to and attracted by opportunities in the criminal economy. Street crimes such as drug trafficking, burglary, prostitution, and car theft become their work (Hagedorn, 1998). The gangs that organize around these activities, while sources of meaningful social relationships, status, and protection for their members, make life in these bleak and isolated inner-city neighborhoods dangerous and problematic.

NOTES AND SUGGESTIONS FOR FURTHER READING

Belkin, Lisa (1999). *Show Me a Hero: A Tale of Murder, Suicide, Race, and Redemption.* New York: Little, Brown.

Bullard, Robert D. (2000). *Dumping in Dixie: Race, Class, and Environmental Quality,* 3rd ed. Boulder, CO: Westview Press.

Burns, Bobby (1988). *Shelter.* Tucson, AZ: University of Arizona Press.

Children's Defense Fund (2001). *The State of America's Children: 2001 Yearbook.* Washington, DC: Children's Defense Fund.

DeOllos, Ione Y. (1997). *On Becoming Homeless: The Shelterization Process for Homeless Families.* Lanham, NY: University Press of America.

Dreier, Peter, and John Atlas (1995). "Housing Policy's Moment of Truth." *American Prospect* 22 (Summer): 68–77.

Eighner, Lars (1993). *Travels with Lizbeth.* New York: St. Martin's Press.

Guzewicz, Tony D. *Down and Out in New York City: Homelessness, A Dishonorable Poverty.* New York: Nova Science Publishers.

Hagedorn, John (1998). *People and Folks: Gangs, Crime and the Underclass in a Rustbelt City,* 2nd ed. Chicago: Lake View Press.

Homeless, Joe (1992). *My Life on the Street: Memoirs of a Faceless Man.* Far Hills, NJ: New Horizon Press.

Liebow, Elliot (1993). *Tell Them Who I Am: The Lives of Homeless Women.* New York: Free Press.

MacLeod, Jay (1995). *Ain't No Makin' It: Aspirations and Attainment in a Low-Income Neighborhood.* Boulder, CO: Westview Press.

McCoy, Frank (2000). "In an Age of Plenty, a Search for Shelter." *U.S. News & World Report* (April 10):28.

Maggi, Laura (2000). "The Squeeze." *The American Prospect* (December 4):28–29.

Massey, Douglas, and Nancy Denton (1993). *American Apartheid: Segregation and the Making of the Underclass.* Cambridge, MA: Harvard University Press.

Parenti, Michael (1978). *Power and the Powerless,* 2nd ed. New York: St. Martin's Press.

Pollard, Kelvin, and Mark Mather (2001). "One in Five U.S. Households Struggle to Afford Rent, Mortgage Payments." *Population Today* 29 (October):3.

The Progressive (2000). "The Housing Crunch." Volume 64 (May):8-10.

Neal, Ruth, and April Allen (1996). *Environmental Justice: An Annotated Bibliography,* Report Series EJRC/CAU-1-96 (Updated June 1998). Atlanta, GA: Environmental Justice Resource Center.

Passaro, Joanne (1996). *The Unequal Homeless: Men on the Streets, Women in Their Place.* New York: Routledge.

Popkin, Susan J., Victoria E. Gwiasda, Lynn M. Olson, Dennis P. Rosenbaum, and Larry Buron (2000). *The Hidden War: Crime and the Tragedy of Public Housing in Chicago.* New Brunswick, NJ: Rutgers University Press.

Ralston, Meredith L. (1996). *"Nobody Wants to Hear Our Truth": Homeless*

Women and Theories of the Welfare State. Westport, CT: Greenwood Press.

Timmer, Doug A. (1988). "Homeless as Deviance: The Ideology of the Shelter." *Free Inquiry in Creative Sociology* 16 (2):163–170.

Timmer, Doug A. (2000). "Urban Problems in the United States." Pp. 139-176 in *Social Problems,* 8th ed., by D. Stanley Eitzen and Maxine Baca Zinn. Boston: Allyn and Bacon.

Timmer, Doug A., D. Stanley Eitzen, and Kathryn D. Talley (1994). *Paths to Homelessness: Extreme Poverty and the Urban Housing Crisis.* Boulder, CO: Westview Press.

Vale, Lawrence J. (2001). *From the Puritans to the Projects: Public Housing and Public Neighbors.* Cambridge, MA: Harvard University Press.

Venkatesh, Sudhir Alladi (2001). *American Project: The Rise and Fall of a Modern Ghetto.* Cambridge, MA: Harvard University Press.

Shelter

Bobby Burns

This selection consists of excerpts from Bobby Burns's diary, his personal account of being homeless on the streets of Tucson, Arizona. He is a college graduate and substitute teacher. Due to a series of events, medical problems, and a drug/alcohol problem, he finds himself without a home, without a job, and in a homeless shelter. Thinking sociologically, what do you learn about shelter life through this author's experience? How would you structure a job program designed to help homeless individuals get back on their feet?

Today I will apply for food stamps for the first time in my life. I arrive at the office at 7:00 A.M., and once the office opens I complete the long application and wait to be called. I feel like I am begging, like a bum off a street corner. The individuals working in this agency treat everyone the same—coldly and impersonally. I sense they see all homeless persons as lazy. I feel angry, humiliated, and ashamed. But as with any agency like this, they have what we need, and we must endure the situation. Long waits are typical in a food stamp office. I wonder if food stamps will bring me a sense of empowerment at the grocery store, being able to buy the things I like in the way of snacks.

Finally, after two hours, my name is called. A young clerk interviews me as she intently reviews my application. She asks me several questions with a bit of a holier-than-thou attitude, then tells me I qualify for $118 in food stamps each month. In two days, I will receive emergency food stamps worth $65. Being homeless has a few perks.

My next stop for the day is the city's Special Services/Transportation offices for a reduced-rate bus card. Getting around any city requires transportation, and I must rely on city buses. Again I sense a cold and insensitive attitude. Many social service agency personnel show little respect or compassion for people who happen to be without an address. Homelessness to me means that I have no home at the moment. It's not a disease a person can catch by simply coming into contact with me.

SOURCE: Bobby Burns, *Shelter* (Tucson, AZ: University of Arizona Press, 1998), pp. 20–21, 30–32.

Before I became homeless, I never took the time to think about or understand who became homeless. I looked through them. What I did know about the homeless I learned from the rhetoric put out by the media or from what I saw with my own eyes. My impression was that their predicament had to be their own fault. Homelessness was their problem, it wasn't mine—until I became part of that population. Now I realize people become homeless for countless reasons. They're not all thugs and bums; some are people who've just fallen on bad times. It seems people get concerned about helping the homeless around the holidays in November and December; they tend to forget the rest of the year. I've learned that homelessness is the same year-round, with some individuals remaining homeless for many years.

My perception of Ivan [shelter worker] is that he is condescending, rude, and arrogant. He talks down to me and another man, and I get the impression he's only half present and his heart isn't in his work. However, he deals with guys like me every day. And who am I? Just another homeless man passing through his doors who says he is trying to find a job. The types of individuals he encounters range from highly employable to those with few or no skills.

Job Connection provides two options to homeless men. A two-week program is for people just passing through town who want to earn enough money to move on. Most jobs pay daily—better known as day labor. One day you might be shoveling up someone's sewer pipes, another day you might be hauling trash to a garbage dump for eight hours for minimum wage. This Job Connection program places no restrictions on clients except that their stay in the shelter is up after two weeks.

The other program, for four to eight weeks, has strings attached. The sooner one finds work, the better. The program offers a résumé service, job skills training, and a message phone. Ivan makes it clear that if I'm not working a full-time job in four weeks, I'm out of the program. If I find work, I can have an additional four weeks in the program.

Each pay day, every client in the program must place 75 percent of his wages in savings, and he can use the remainder any way he wants. Each person is responsible for getting a money order in his name and handing it over to Ivan. Earnings are returned at the end of the program. A person can get ahead and land back on his feet if he sticks to the plan. I believe I can if I don't drink. I have no excuse not to succeed. My meals and rent are free. Finding a job seems a minor concern. I agree to the terms of the contract and sign on the dotted line.

My next priority is to go back to the county Health Department for my TB results. A nurse looks at my left arm and says my test is negative. She hands me a card showing that I am TB-free, which I will have to show the shelter. It's 11:30 A.M. and I decide to eat lunch at a downtown greasy spoon. I order a bowl of soup, a half-sandwich of tuna, and a cold glass of iced tea. I sit on the café patio as the lunch crowd comes and goes. The weather is sunny and breezy.

I later return to the R & R office to see if my food stamps have arrived. A volunteer hands me my mail. A letter from the Arizona Department of Economic Security informs me that I will receive my food stamps within five days. Leaving the office, I see a man trying to sell his food stamps to another man for cash. This sort of thing happens all the time. With cash you have more latitude. With food stamps you face certain restrictions. I walk downtown pondering what I should be doing with myself. Idle time is what gets me in trouble with booze and drugs, this I know.

Time has gone by fast today. I walk a few blocks from downtown to catch the shelter bus. As I approach the pickup point, I notice about twenty men waiting for the bus, some men I have never seen before. A few are talking, and the rest stand in silence as if they have nothing in common with one another. The bus ride is mellow as the big white bus passes through the downtown.

Uncertainty enters my mind as I size up the other men, who look as unsure and wary as I. As the bus arrives at the shelter, a manager waits at the front door with a clipboard in his hand. "Repot to the second door on your right if it's your first time in the shelter," he yells.

Those of us who already have beds give our names and bunk numbers by rote. All of my possessions are shuffled onto and under my bed. I feel discouraged as I look around the cluttered barracks. Homelessness is undeniable. Still, I have problems in accepting that I have fallen this low. The oppressive environment on a daily basis, the day-to-day grind, gets to me. Every day I face the constant noise, the bothersome loudspeaker, the rules, the schedule, strange people in and out. But if I can put up with navy boot camp, then I can put up with the shelter.

Bureaucracy and Rules as Punishment

The Roofless Women's Action Research Mobilization and Marie Kennedy

This selection points to the negatives associated with life in homeless shelters. The criticisms are from women who have lived in shelters in the Boston area. Homeless shelters have, by necessity, rules for the smooth functioning of the organization and its clients. At what point does the bureaucracy interfere with personal freedom and responsibility? Think about the different theoretical perspectives on poverty. If you believed in the individual/cultural perspective, how would you structure a homeless shelter? If you believed in the structural perspective, how would your shelter be structured?

Once homeless, most women face a system that strips them of their self-esteem, limits their ability to parent their children, and, in general, delivers services in a punitive way that encourages self-blame. For homeless women with children, the abuse frequently starts with the welfare office. Most homeless shelters are funded directly by the DPW, and, to get in, you have to be approved for Emergency Assistance. Having fled for her life back to her home town of Boston, Barbara was told by the DPW housing search worker, "There's nothing here for you. The best we can do is pay your way back to the South where you came from." For *Barbara* this was like "a slap in the face."

Women often have to prove to Welfare that they are, indeed, homeless before they can get Emergency Assistance. Knowing there was nothing she could do to prevent her eviction, *Joan* went to Welfare to be placed with her two children in a shelter:

> . . . they want to give you a hard way to go and tell you, "Well, you know, we're not going to do anything for you." And I got really mad and I said, "I've never been on AFDC, I've

SOURCE: Roofless Women's Action Research Mobilization Researchers: Deborah Clarke, Delores Dell, Brenda Farrell, Deborah Gray, Betsy Santiago, and Tesley Utley, edited and narrated by Marie Kennedy, "A Hole in My Soul: Experiences of Homeless Women," *For Crying Out Loud: Women's Poverty in the United States,* Diane Dujon and Ann Withorn (eds.), (Boston: South End Press, 1996), pp. 47–50. Used by permission of the publisher.

> worked all my life and it's my right to have housing." So, they made me wait until the day I got evicted at 4 o'clock before they told me where I was going to go.

Many have their families pressured by Welfare to take them in, even if, as in *Janice's* case, there has been little contact for years.

> When you apply to get into the shelter, they call your family to see if they'll take you back—your mother, your father, your brothers, and your sisters. My mother had to sign a release saying that I could not live with her. I hadn't had any communication with some members of my family, and then to have the Department of Public Welfare call and say, "Could she come to live with you?" Like, "I haven't talked to her in seven years." Click. That's disgusting. It all gets back to that they think you're trying to deceive them. There's this perception that if you have to apply for welfare, then you are a crook.

Of course, many shelters are not much better. When *Sandra* went to her first Boston shelter, the only identification she had was her passport. After flipping through it, the intake worker sarcastically commented, "You've been to all these places and you're still homeless?" Upon learning that *Sandra* had also been to college, the worker continued in her sarcastic vein, "Four years of college and you still couldn't manage to not become homeless!" Then she encouraged Sandra to return to New York, rather than be a burden on Massachusetts.

Once in a shelter, women face rules and staff behavior that are frequently demeaning and that undermine their efforts to get back on their feet. Shelters for women without children are infamous for treating clients like nonpeople and preventing them from having any sense of security or community. Most shelters for single adults kick everyone out by eight in the morning, even on the coldest days. Some shelters require that clients be back too early to be able to work a normal job. Delores Dell describes one example:

> There's this shelter in Somerville that drops the women off in Central Square [a commercial area in the next city] at 7 or 7:30 in the morning and they have to be back at a certain location at 4 p.m. in the evening to be shipped out to Somerville. If you miss that bus, you don't have a shelter to stay at that night. Also, there are 21 beds there and if there are 22 women, they have to have a lottery and whoever gets the "X"—the dreaded "X"—they give them tokens to find another shelter to stay in for that night. It doesn't matter if you've been staying there for six months every night; if you draw the "X," you're shelterless for the night.

And, where do women go on a cold day when they've been kicked out of the shelter? *Sandra* describes what she faced before a women's drop-in center opened:

> . . . you go someplace to keep warm, and that was the drop-in at [large shelter]. The minute you walk in, a hundred people run up to you and they make all kinds of weird propositions. It's a scary place even for someone who has street smarts as I do; it's not a safe place for women. People are attacked from time to time. Drug use is rampant in the bathrooms and things that I'm not even going to mention occur in little nooks and crannies.

Although many shelters talk about the empowerment of their guests, too often it does not mean trusting guests to know what's best for them, Delores Dell points out, but rather things like "letting the guests do their own laundry." Meanwhile, shelter rules constrain behavior, and the threat of losing shelter is used to keep people in line. For *Tess,* this meant that she couldn't even enjoy a sociable weekend away with friends:

> You'd be there every night, but if someone invited you to stay over on the weekend, the shelter said that if they have a room for you on a weekend, then they must have room for you all the time, so, if you go, don't bother coming back.

The culture in rule-bound shelters encourages even well-intentioned staff to treat clients like bad children. *Sandra* related an all-too-typical incident:

> One of the workers—a nice person—got into a disagreement with one of the other women staying in the shelter and said something like, "I'm not homeless. I have a place to go to. And, if you give me a hard time, I'll put you out right now." The woman was just trying to explain her side of the story and the staff person said, "That's it, I'm not having any more. Get your things and get out. You've got two minutes or I'll call the cops."

Many shelters seem to encourage a culture of dependency. Even if they don't evict a woman because she shows some independence, they can make her life more difficult. *Janice* never felt welcome in her shelter and felt that the threat of eviction was always there:

> When they were doing intake, they told me they didn't even want me there and that they were taking me only because they were short on funding and they needed the money from AFDC—the $2,000 or $2,200 a month they got to keep me there. It was like a constant threat. If I didn't do what they told me, I wouldn't be able to put my kids in the bed that night. And, they told me what my needs were: "This is what you need to do." And, I said, "That's not what I need to do." Their focus was to teach me how to live on welfare. They had budgeting classes . . . I already knew how to budget. I didn't want to live on welfare. My goal in life was not to get a $200-a-month apartment and live on welfare. My goal was to get back on my feet, go back to work, and start my life over. But, because I didn't go to the classes, because I wasn't following what they wanted me to do, I was an outcast. I was out trying to find housing and that wasn't their plan. The whole system is set up not to treat you as an individual, but as one of a mass. How can you feel any self-esteem when you're cattle? You're basically herded through this system and they have this gate you're supposed to go through and end up here; that's what they're projecting for you, but that may not be where you want to go.

The routine invasion of privacy is another way that homeless people are treated as objects or nonentities. Whether in a single or family shelter, your life is an open book. The interest of staff members in knowing personal details can be a way of exercising power, a type of voyeurism that is not connected to job requirements. Staff in *Janice's* shelter felt free to ask detailed questions about the spousal abuse she had suffered. She feared she would be kicked out of the shelter if she didn't answer. As she put it:

> They're discussing me and my violence and all that stuff and I'm thinking, they have no right. Except, they felt they did. And that whole power thing—it was like a constant threat: "If you don't do what we say, tell us what we ask, we're going to take your shelter away."

The Residents of Rockwell Gardens

Susan J. Popkin, Victoria E. Gwiasda, Lynn M. Olson,
Dennis P. Rosenbaum, and Larry Buron

In The Hidden War: Crime and the Tragedy of Public Housing in Chicago *the authors surveyed residents and observed life in three Chicago housing projects over a period of four years. The result is a portrait of life in some of the poorest neighborhoods in the United States. This selection is the perspective of one of the residents of Rockwell Gardens. Analyze Dawn's housing situation/neighborhood from the individual/cultural and structural perspectives. What are some possible solutions to the problems of Rockwell Gardens?*

DAWN'S STORY

Dawn is a long-time resident of Rockwell Gardens—the development widely considered the most dangerous of the Chicago Housing Authority's (CHA) high-rise projects. She is a thoughtful, articulate woman in her early thirties and has lived in Rockwell since she was fourteen. Dawn became pregnant as the result of a rape when she was sixteen; after the birth of her second child two years later, she dropped out of school to care for her children. Dawn's mother could not help her with child care because she had young children of her own, and Dawn could not find another sitter she felt she could trust.

Her children are now teenagers, and she spends most of her time and energy protecting them from what she calls "the pressure" of the surrounding environment: young men pressure her daughter for sex, and gang members try to recruit her son. For years, she has monitored her children closely by taking them to parks outside the development to play and keeping them indoors as much as possible. "They mostly sit on the porch in the building. . . . [They go out] with me. . . . Sometime they go to the store and back, but I don't really let them hang in front of the buildings and stuff. If they go to a playground, they go in the Maplewood Courts playground [an adjacent, low-rise development] because they don't shoot over there. Or across the [freeway] bridge to the big park." Although her son sometimes goes to the store or participates in after-school programs, Dawn says her daughter no longer goes outside alone.

Dawn has been on welfare since the birth of her first child. She says she always wanted to be a nurse when she was growing up, but "her children held her back." Although she did eventually get her GED, her only work experience has been an "off-the-books" job as a housekeeper for a few years. She desperately wants a better life for herself and her children, but she lacks practical ideas about how to improve her situation.

Living in CHA housing has taken a toll on Dawn and her family. Everyone she knows lives in public housing: her entire extended family lives in CHA developments all around the city. She says that no one in her family has "made it," that all the developments are "just the same." Three members of her extended family have died because of gang violence, one as a result of a domestic

SOURCE: Susan J. Popkin, Victoria E. Gwiasda, Lynn M. Olson, Dennis P. Rosenbaum, and Larry Buron. *The Hidden War: Crime and Tragedy of Public Housing in Chicago* (Piscataway, NJ: Rutgers University Press, 2000), pp. 39–43. Used by permission of the publisher.

dispute. Several, addicted to drugs, have lost custody of their children. In April 1996, she told us that, since January, she had already been to three funerals for people from Rockwell, which she said was "typical." Dawn is often depressed, but says she had no one to talk to about her feelings because "people in the projects, they don't want to hear your problems. They tell you 'I have problems of my own.'"

Throughout the two years we interviewed her, Dawn offered many insights into the challenges of life in Rockwell. Although her greatest concerns were drug trafficking and violence, the daily problems that she found most frustrating were poor maintenance and unresponsive housing authority management. For example, in 1994, Dawn said she had been coping with severe plumbing problems for two years:

> My tub water don't shut off—the hot water . . . and the kitchen won't shut off. . . . Well, they just fixed my sink not too long ago because they had to take the whole pipe out of the wall because it was full of that greasy, gunky stuff and the water wouldn't flow through and it kept running out, running out. That took them two years to get up and fix that. I had to wash my dishes in my tub every day. . . . [In the bathroom] my walls were white, [now] they yellowish-brownish from the hot water. I have to close the door at night to keep from going crazy hearing that water! . . . . I been waiting six months for them to fix that.

Vandalism compounded the maintenance problems and, Dawn felt, made her building an even more frightening place to live.

> Some people may set their garbage . . . in front of the incinerator, which they not supposed to do that, they supposed to take it downstairs to the main floor [because the incinerators don't work], but they don't. Either the kids will set it on fire, or the big boys will come and just throw it down the steps, and it's really dark, we can't see! You know . . . I bust my head, bumped into the light room door, it being dark. . . . And then when they do put bulbs in every day, the guys will like take sticks and break the bulbs or take 'em out and make it dark. Seem like they like it dark.

In Dawn's view, drugs, particularly cocaine, have completely devastated the Rockwell community. She spoke poignantly of seeing friends who were once "strong women" wasting away because of their crack addictions. She said she could no longer let some people she knows into her apartment because they would try to steal whatever they could to support their habit.

> Man, the rock's [crack] been here seem like forever. Because when I used to come from the grocery store and I be bringing my groceries in, the elevators be broke, and we used to have to carry our stuff up the stairs . . . and there's people in the hallway smoking crack with . . . they was actually using t.v. antennas. . . . People come to your door and—they say, "Can I use your bathroom," I always tell them no, and I know them, but I tell them no because they wanna go in there and break off a piece of your t.v. antenna.

Dawn said she had never used drugs herself, and she stayed away from "project men," whom she described as pressuring their girlfriends to become users so they could use their welfare money to buy drugs.

Dawn was particularly concerned about how the disintegration of the community affected the children of

Rockwell. She agonized over the children she saw being abused or neglected by mothers addicted to cocaine, crack, and heroin. She told the following story to illustrate the extremes to which addicts went to feed their addiction: "Them women just go crazy for that stuff, they do anything. . . . Anything. They don't care. One lady sold her baby for some rock. . . . To the man, it was a guy who sold cocaine over there, she . . . gave him her baby, she told him to hold my baby until I come back, I'm gonna go get the money, give me the rocks."

Dawn used to try to intervene when she saw hungry or abandoned children in her building. When she could afford to, she fed them herself. But when she found children who were unsupervised—for example, a five-year-old taking care of a four-month-old—she would sometimes call the Department of Children and Family Services to report the situation. When the social workers came, she would help getting the children dressed and ready to be taken away. She also used to call the police to report shootings or other crimes. Eventually, her family convinced her that it was too dangerous to get involved, a position that causes her much pain. She spoke of her torment over having kept silent after she witnessed a young neighbor using a gun:

> When you in the projects, you do a lot of things, you see a lot of things, but you know you don't wanna say nothing because it can get you hurt . . . but it be on your conscience, and it drives me crazy when I can't say nothing. . . . I see this little boy, he's about twelve years old. He's shooting, I see him shooting at the other [kids]. And I'm looking at this, and I know his mother and everything. Everybody telling me, "No, don't say nothing, don't tell his mother." And now he's dead, the little boy is dead now, and it made me feel if I had a told his mother, maybe he'd still be here.

Dawn said she had high hopes when the CHA first started the sweeps in Rockwell in 1988; she believed they would help to control the gangs and drug trafficking. For a time, she felt that conditions did improve a little, but, by the time we met her in 1994, she told us that all the good effects had disappeared. However, she was hopeful that the CHA's latest effort—hiring a company owned by the Nation of Islam to provide security and manage the development—would help. Also, she reported that things were getting calmer following a gang truce in Rockwell.

But from 1994 to 1996, Dawn became increasingly disillusioned and unhappy. In her view, the "Muslims" proved no better than any other security guards who had patrolled Rockwell over the years. The new management cleaned up the buildings, but she still had serious maintenance problems in her apartment. Drug abuse was still rampant, and the gangs still dominated the development. Sometimes she felt that things had improved for a little while because of a gang truce or the arrest of a particularly powerful gang leader, but the gang war between the Disciples and the Vice Lords always "flared up" again after a month or two.

In spring 1996, Dawn finally moved out of Rockwell Gardens to another, smaller CHA development. A friend of hers had moved there and told that it was "nice." Dawn was dismayed to find that, although the complex was cleaner than Rockwell and the apartments were better, she still faced many of the same problems: "I only been there not long at all and they been shooting, I been hearing gun shots, one boy done got killed, right out by the building in the back."

6

The Welfare System

From the time of the Great Depression to 1996, the United States provided a minimal safety net for those in need. The New Deal under President Roosevelt and the Great Society under President Johnson created the minimum wage, federal aid to education, health and nutrition programs, food stamps, energy assistance, subsidized housing, and Aid to Families with Dependent Children (Moen and Forest, 1999:644–647). These minimal but helpful programs have gradually been dismantled, beginning with President Nixon. This dismantling quickened appreciably in 1996 when the federal government made welfare assistance to families temporary and withdrew $55 billion of federal aid to the poor.

THE PERSONAL RESPONSIBILITY AND WORK OPPORTUNITY RECONCILIATION ACT OF 1996

In 1996 the Republican-dominated Congress and a middle-of-the-road Democratic president passed a sweeping welfare law that ended the 61-year-old safety net for the poor. The major provisions of this law (as later amended) included the following:

- Through federal block grants states were given a fixed amount of money for welfare and considerable flexibility in how to spend it.

- The law insisted on work. The states were required to demand that parents work within two years of receiving cash assistance, although states were free to shorten the period before welfare recipients must work.
- The law mandated a five-year lifetime limit on the receipt of assistance; states can reduce this limit if they wish.
- The law required that unmarried teen parents must live at home or in another adult-supervised setting and attend school to receive welfare assistance.
- Various federal assistance programs targeted for the poor were cut by $54.5 billion over six years. Included in these budget cuts were $24 billion from the food stamp program, $7 billion from the children's portion of the Supplemental Security Income program, $3 billion for child nutrition, and $2.5 billion for social services. Cuts were also made by tightening the qualifying criteria for being defined as a disabled child.
- The welfare law denied a broad range of public benefits to legal immigrants.
- The federal money given to the states was capped at $16.4 billion annually. This is significant because it means that there is no adjustment for inflation, business downturns, and population growth.

In sum, this new welfare legislation ended the entitlement which guaranteed that states must give help to all needy families with children. Now assistance for poor families is temporary (Aid to Families with Dependent Children—AFDC—was replaced with Temporary Assistance for Needy Families—TANF), with parents required to work.

A question arising from the requirement that all former AFDC mothers work is whether a single mother is "able" to work (McLarin, 1995). Traditionally, a single mother was not considered able to do so. AFDC was created in 1935 with the goal of keeping women at home with their children. The 1996 legislation changed that, forcing poor women with children to work—without the training, without the jobs, and without the childcare. Through twisted logic, the same conservative politicians who want poor mothers to work want middle-class mothers to give up their jobs because a stay-at-home mother is positive for children.

Two additional problems arise with forcing single mothers to work. First, most of the available work for women with little education and few skills pays at or slightly above the minimum wage, a wage that gives a full-time worker an annual income that is several thousand dollars *below* the poverty line. Work also entails the additional costs of child care, transportation, and clothing. Second, when economic times are good, there may be enough jobs available for those no longer eligible for welfare, but what happens to these women and their children when there is an economic downturn and the last hired are the first fired? This question has been answered with the economic downturn beginning in early 2000, the subsequent recession, and the terrorist attacks on the World

Trade Center and the Pentagon on September 11, 2001, which accelerated unemployment in certain industries. In October 2001 alone, 415,000 jobs were lost, many in the very industries dominated by temporary and part-time workers and low-wage jobs where former welfare recipients found jobs when the economy was booming. With the economic slump, unemployment in general has risen sharply, with the last hired (many of them former welfare recipients), the first fired. Moreover, only 40 percent of those workers formerly on welfare have worked long enough to receive unemployment insurance. These newly unemployed are separated from jobs with much less of a safety net than just a few years ago, caused by reductions by the federal government and budget tightening by state governments. The result is that some former welfare recipients are poorer than ever, with their housing arrangements less secure, and enough food ever more problematic. Regarding the latter, Second Harvest, a nonprofit supplier of much of the nation's emergency food aid, estimated that 23.2 million Americans received emergency food aid in 2001, 40 percent of them children, up from 21.4 million in 1997 (reported in Bernstein, 2001).

Social policy is about design, involving the setting of societal goals and determining the means to achieve them. Should we create and invest in policies and programs that protect citizens from poverty, unemployment, inferior educations, and inadequate medical care, or should the market economy sort people into winners, players, and losers based on their abilities and efforts? Decision makers in the United States have opted to eliminate most of the safety net for those on the economic margins, especially single women with children. Are these policy makers on the right track? Is this design the solution to poverty, or should the United States change its design?

The voices in the following selections are from people who have lived on welfare. After reading these selections, how would you answer these questions? Is welfare a good policy? If so, who should receive it? How should the welfare system be changed?

NOTES AND SUGGESTIONS FOR FURTHER READING

Bernstein, Aaron (2001). "Already, A Crush at the Soup Kitchens." *Business Week* (November 26):74.

Blumenberg, Evelyn (2000). "Moving Welfare Participants to Work: Women, Transportation, and Welfare Reform." *Affilia* 15 (Summer):259–276.

Edelman, Peter (1997). "The Worst Thing Bill Clinton Has Done." *Atlantic Monthly* (March):43–58.

Eitzen, D. Stanley, and Maxine Baca Zinn (2003). *Social Problems,* 9th edition. Boston: Allyn and Bacon.

Ellwood, David T. (2000). "Anti-Poverty Policy for Families in the Next Century: From Welfare to Work—and Worries." *Journal of Economic Perspectives* 14:187–198.

Foust, Dean (2000). "Easy Money." *Business Week* (April 24):107–114.

Funiciello, Theresa (1993). *Tyranny of Kindness: Dismantling the Welfare System to End Poverty in America.* New York: Atlantic Monthly Press.

Gans, Herbert J. (1995). *The War Against the Poor: The Underclass and Antipoverty Policy*. New York: Basic Books.

Gilbert, Neil (1998). *Dimensions of Social Welfare Policy,* 4th ed. Boston, MA: Allyn and Bacon.

Handler, Joel F. (1995). *The Poverty of Welfare Reform*. New Haven, CT: Yale University Press.

Harris, Kathleen Mullan (1996). "The Reforms Will Hurt, Not Help Poor Women and Children." *The Chronicle of Higher Education* (October 4):37.

McLarin, Kimberly J. (1995). "For the Poor, Defining Who Deserves What." *New York Times* (September 17):4E.

Moen, Phyllis, and Kay B. Forest (1999). "Strengthening Families: Policy Issues for the Twenty-First Century." Pp. 633–663 in *Handbook of Marriage and the Family,* 2nd ed., Marvin B. Sussman, Suzanne K. Steinmetz, and Gary W. Peterson (eds.). New York: Plenum Press.

O'Connor, Alice (2001). *Poverty Knowledge: Social Science, Social Policy, and the Poor in Twentieth-Century U.S. History*. Princeton, NJ: Princeton University Press.

Polakow, Valerie (1999). "Savage Distributions: Welfare Myths and Daily Lives." Pp. 241–262 in *A New Introduction to Poverty: The Role of Race, Power, and Politics.* Louis Kushnick and James Jennings (eds.). New York: New York University Press.

Peck, Jamie (2001). *Workfare States*. New York: The Guilford Press.

Seccombe, Karen (1999). *"So You Think I Drive a Cadillac?"* Boston: Allyn and Bacon.

Working Your Fingers to the Bone

Marion Graham

Marion Graham raised five children while enduring an abusive husband and then using welfare and low-wage jobs to survive. She was a union steward and a cofounder of ARMS, the student welfare rights chapter at the University of Massachusetts/Boston. As you read her essay, think about the poverty line, its definition, and its use as a measurement for public assistance. Is there a better way to measure poverty and to determine qualifications for assistance? What does this author mean by "invisible poverty"? Is this author demonstrating the sociological perspective in this essay?

I was born in Boston, the youngest of a four-girl family. When I was 13, my father's firm went bankrupt and he found himself out of a job. My family were so ashamed they wouldn't tell their friends that they had to move from the suburbs to a three-decker in Boston, even though they weren't really poor. Now I know a lot of people who would love to move out of the projects into a three-decker.

In 1960 I got married. I left my good job with the telephone company when I was pregnant with my first child, in 1961. I really looked forward to being home with my children. Nobody worked that I knew. During the '60s I was always pregnant when everybody was out rebelling against everything. I was too pregnant to rebel, so I have to rebel now! I have five kids. They are now 35, 32, 31, 30, and 27.

When I saw how my marriage was disintegrating, I did work at home for

SOURCE: Marion Graham, "Working Your Fingers to the Bone," *For Crying Out Loud: Women's Poverty in the United States,* Diane Dujon and Ann Withorn (eds.), (Boston: South End Press, 1996), pp. 151–153. Used by permission of the publisher.

marketing research companies, and I did typing for college students, just to try to make money. I had planned for two or three years to get a divorce before I did. But I never had the money to do it. I knew I had to have a job in order to save the money to go to a lawyer. I couldn't leave the kids; there was no daycare. Finally, I had to go on welfare because my husband did not pay enough support, and sometimes he did not pay at all.

When I started working full-time again, I thought it was going to be wonderful, that I wasn't going to be poor anymore. I was going to be away from the bureaucracy; they couldn't call me in anytime they wanted to. Even then, though, I still earned so little that I was eligible for a housing subsidy, Medicaid, and food stamps. I remember at the time being ashamed to let people where I worked know that I was poor enough for food stamps. And I hated that.

About three years ago they came out with some new "poverty line," that's what they called it, and they decided I earned too much for most of those other benefits. I only grossed something like $8,600 at the time, but it was still too much for them, so they cut me off. I still had the same needs I had before, but suddenly I was no longer poor. I guess I was supposed to be proud.

Since then I got a raise, so I thought things would be O.K., but then I lost my housing subsidy because I earned too much. I ended up having to take an apartment that cost exactly seven times what I had paid with a subsidy. My rent came to half of my net pay. Just the rent. After that sometimes I would get to work but not be able to pay for lunch. I had my subway tokens, but no money to eat. Every week they take something different out of my salary. Life insurance, disability insurance, union dues, retirement benefits are all good, but you don't get much to live on. Now, finally, I can buy *Woman's Day* and *Family Circle* magazines—that used to be my dream.

How you dress for work, the hours and flexibility, transportation, whether you can bring a lunch—all these nitty-gritty things make a big difference in how you can live on a low salary. You just cannot afford to take some jobs even if they sound interesting because you have to spend too much money on clothes or transportation. It's sad. I couldn't afford to work at a place where I had to dress up, for instance. You can't "dress for success" on a secretarial salary. It's an invisible poverty.

You think you're not poor because you are working, so you don't even ask for the information about benefits you need. And nobody tells you that you might be eligible, because they think you are working and all set. Also, it is even harder to ask for things from your family, because if you are working you should have the money. I feel bad, though, that I don't have more chance to help my family. I can't afford it, even though I'm working.

The average pay in my union of clerical and hospital workers is not much above poverty for a woman who is trying to raise a family. When people need childcare they have to pay for it, and they can't afford it. I don't need childcare, but the health insurance, which I had to wait two years for, costs a lot. And I couldn't get it for my son, who has suffered from juvenile diabetes since he was very young.

Not having money affects everything about how you feel. I used to feel lousy about myself. I thought I was supposed to be set, to have a slice of the American pie. Now I was a big person and I worked and everything, and I was supposed to get there. But instead I found myself just with a job and no money. When I reached 40 I was so depressed.

Now I have learned that I am not alone, that it is not my fault. The average secretary around here is just over the line for many benefits, but we still have expenses we can't meet. That makes some women, who don't understand how it works, take it out on women on welfare. They blame them for getting something they can't have, instead of blaming the rules, which keep them from getting anything. Some secretaries may think, "I am better off than they are," instead of seeing how we have similar problems. But they are afraid of the label that would be put on them if they identified with welfare. I don't do that because I have been there and I know both, and I know that none of it is good. It's bad to be on welfare, and it is bad to be working and have no money.

As secretaries here we have worked hard to do things together so that we can know each other as people, because at work we are all separated in our individual little offices. We go on picnics together, or to dinner, just to get to know each other. Although they don't pay us much, they act like we can never be absent or the world will fall apart, so we have to cover for each other, and we can't do that if we don't know each other. I keep saying, "If we are so important, why don't they pay us more?" But they don't, so we have to help each other.

Welfare As They Know It

Plaintiffs in *Capers v. Giuliani*

The following selection consists of testimony in a class-action suit brought on behalf of New York City workfare participants. Workfare participants are individuals who are required to work in order to receive welfare benefits. This particular class-action suit involved those individuals who were assigned to the Departments of Sanitation and Transportation in New York City, and their testimonies document the hazardous working conditions they encounter. As a result of their testimonies, a State Supreme Court Justice ruled that the city was obligated to improve the working conditions of its 5,000 workfare participants. Think about welfare reform that requires recipients to work to receive their benefits. What problems with this system are brought to light by the following testimonies? Are any working conditions better than none at all? Think about workfare and women with small children at home. What potential problems do you foresee for these women?

SOURCE: Plaintiffs in *Capers v. Giuliani*, "Welfare As They Know It," *Harper's Magazine* 295 (November 1997), pp. 24–26.

Tamika Capers, Age Twenty-One

Since February 26, 1997, I have been cleaning highways and the areas next to highways in the Work Experience Program (WEP) for the New York City Department of Transportation. The work consists of raking, sweeping, picking up garbage, cutting tall grass by hand with a long-handled Weedwacker, and clipping tree branches overhead.

While we clean the highways, cars, trucks, and tractor-trailers whiz by only a few feet away. Cones are put up to protect us, but they are spaced so far apart that the cars ignore them and swerve inside when the traffic is heavy. Last month, a car went out of control and ran up on the grass, heading right for two of my co-workers, who had to dash out of the way to avoid being hit. When the car returned to the highway, it stopped, causing a three-car pileup.

We have no access to a toilet either in the parking lot where we are picked up each morning or out on the highway. If we need to urinate or move our bowels, we have to squat behind a tree or bush or ask one of our co-workers to hold up a plastic bag to shield us from the passing cars. We have not been given insect repellent, and I am afraid to relieve myself outdoors, because I will be exposed to the many biting insects flying and crawling around us. My stomach cramps from holding my urine. During my menstrual period, there is no place to go to change my pad. I have to wait until the end of our shift, and by then my clothes are soaked with blood.

Anastacio Serrano, Age Forty-Four

On March 10, 1997, I began my WEP assignment on a cleaning crew of four people. We sweep twenty to thirty blocks of street per day in the Williamsburg section of Brooklyn. Sometimes our group travels in a van behind a garbage truck. When we see trash, the van stops and the four of us get out of the van, sweep up the mess, and put it into the garbage truck. The supervisor sits in the van the entire time.

I keep a log of my daily activities in WEP. On June 18, while riding in the van, we came across two dead cats and two dead dogs. They had been dumped by the side of the road. Because I have no gloves, I had to pick them up with my bare hands. The animals had been run over by automobiles and were oozing blood and entrails. When I picked up the animals with my bare hands to throw them into the garbage truck, the guts splattered on my shoes and pants. My co-worker vomited. My supervisor, sitting in the van, said nothing. I have seen other people who were terminated by my supervisor for refusing to pick up things, and I was afraid that if I refused to leave the van or left the carcasses in the gutter I would be terminated also.

Because I have no dust mask or eye protection, I suffer from the dust that blows up in my face while I am sweeping. I have glaucoma, which requires me to use pilocarpine hydrochloride drops four times a day. The dust gets in my eyes and dries up the solutions. I can hardly see by lunchtime. I have only one eye that I can see from now, and I am worried that I may eventually lose my sight completely.

I have trouble sleeping for the first time in my life. I toss and turn at night because of the pain in my eyes and back and from the stress of having to endure the daily humiliation of this program.

Mery Mejia, Age Thirty-Seven

Since February 1997 I have been working on a crew that cleans the ramps and service roads near the highways. I cut weeds and grass all morning by swinging a long wooden stick with a flat metal blade. The tool is heavy and hurts my shoulders after a few minutes. I'm about five feet tall, and some of the grass is taller than I am.

While working, I see rats everywhere along the roads. They don't bite us, because we scream and run away. They are about eight inches long, with long, stringy tails. They are sometimes right next to us. There are also all sorts of plants, maybe poison ivy, that give me rashes. But the supervisors give us no work clothes, either in the winter or the summer. The only things we get for protection are an orange vest, a pair of cotton gloves, and a hard hat. The gloves and the hard hat are filthy, but I am not allowed to take anything home to wash it. They gave out boots for one week in February, but they were all size 12 and did not fit me.

Because there is no place to go to the bathroom, I do not drink water. This makes me feel like I am suffocating. When I work I feel nauseated and lightheaded, and I get terrible headaches. When I finally do go to the bathroom my urine is very dark brown, which makes me concerned about my kidneys. One of the other women in my crew fainted from the heat, but we caught her before her head hit the ground.

When I work, the dust gets on my clothes, in my eyes, and in my nose. When I blow my nose during work the mucus is dark brown because I don't have a dust mask. I change my clothes when I get home, but I have to wash them with all the other family clothes, because I cannot afford to wash them separately. Sometimes my clothes are black with dirt.

My health and my spirits have been worn down by this program.

Omar Torres, Age Thirty-Eight

I never received an orientation or any type of preparation before beginning my WEP assignment in June of 1995. Not once on the job have I been advised what type of clothing to wear. Once I was assigned to clean graffiti off of a fleet of fifty garbage trucks. I was never provided with goggles, though I used heavy equipment to spray graffiti cleaner onto the garbage trucks. I have scars on my legs where I was burned by splashing graffiti-cleaning fluid.

Before beginning any particular job for the day, the supervisors distribute the orange safety vests that WEP workers are required to wear. The vests at the garage where I am based are kept in a cat-litter box on the floor. I am not allowed to take the vest home to clean it, nor am I allowed to keep one for my individual use.

I cannot take my lunch to work, because there is nowhere to store it. I am not allowed to keep any belongings, including food, at the garage. I cannot carry it with me while I am cleaning the streets, because there is no clean place to keep it. Some people tie their lunch to their garbage barrel handle.

My supervisor refuses to listen to my problems. When I verbally challenge his opinions about working conditions, he threatens to terminate me. I am very frustrated by the unjust way I am treated. The supervisors are like overseers on a plantation. They constantly threaten the WEP workers with sanctions and termination. They try to brainwash you into thinking that no lawyer can help you, and tell us not to talk to them. Daily I have to think about whether a store will let me use the bathroom. It is constant stress. I feel like we are treated as if we were not human beings. I should not have to go through this just to work.

Sylvia Ruff, Age Fifty-Seven

Since March 1997 I have worked for the New York City Department of Sanitation, sweeping up streets and picking up garbage. The garbage includes broken glass, nails, syringes, needles, used diapers, used condoms, used tampons, and dead rats. We also encounter a lot of half-empty containers with strange-smelling liquids in them and a lot of dirty, discarded clothing, which may be infested with germs.

There is nowhere to wash your hands before lunch, and my hands are often dirty from picking up garbage. I always worry about germs and often I do not eat. And now that the weather is hot, I get terribly thirsty on the job, but there is no bathroom to use and I am afraid of having to urinate with nowhere to go. Even if I dared to drink, there is nowhere at the work site to get water. The supervisors threaten to sanction us when we leave our route to go to a restaurant and get water.

One day in early June when it rained, I was given a lightweight orange poncho to wear, which left my forearms and my legs exposed. The hood was designed so that I could not see to my right or left. Since I need to see oncoming traffic, I could not use the hood. It was pouring rain, and we were out for hours, standing in puddles as we tried to sweep. I got soaked through. I had to wring out the sleeves on my own jacket, and when I got home my feet looked like prunes.

The fear of humiliation I face because I have nowhere to relieve myself is degrading. The exposure to chemicals and waste products is unhealthy for me. I am worried about bringing my polluted clothing into the house and endangering my son's health as well as my own. Not knowing what I am being exposed to makes me constantly anxious, and it is a daily struggle to overcome this. The constant threats by the supervisors to sanction us for leaving our route and the inhumane working conditions make it seem like we're on a chain gang instead of in a work-experience program. I have been told—and I believe—that this work will not lead to a job. As long as we are forced to do this dead-end work, we at least need access to bathrooms and clean water, adequate personal protective clothing and equipment, and meaningful training about the hazards we face.

Welfare

Rita Henley Jensen

Here, former welfare recipient Rita Henley Jensen talks about her experiences and feelings toward the welfare system. She is now an investigative journalist and received welfare while she was earning her bachelor's degree. As you read her essay, think about this author's experiences and current situation. Do you think that working on a college degree should count as a job or job training so that women can be on welfare and go to school? Why/why not? From her essay, find some specific examples that demonstrate the structural perspective. PROWRA (welfare reform legislation) was passed by President Clinton one year after this was written. How would Jensen feel about the reform?

I am a woman. A white woman, once poor but no longer. I am not lazy, never was. I am a middle-aged woman, with two grown daughters. I was a welfare mother, one of those women society considers less than nothing.

I should have applied for Aid to Families with Dependent Children when I was 18 years old, pregnant with my first child, and living with a boyfriend who slapped me around. But I didn't.

I remember talking it over at the time with a friend. I lived in the neighborhood that surrounds the vast Columbus campus of Ohio State University. Students, faculty, hangers-on, hippies, runaways, and recent emigres from Kentucky lived side by side in the area's relatively inexpensive housing. I was a runaway.

On a particularly warm midsummer's day, I stood on High Street, directly across from the campus' main entrance, with an older, more sophisticated friend, wondering what to do with my life. With my swollen belly, all hope of my being able to cross the street and enroll in the university had evaporated. Now, I was seeking advice about how merely to survive, to escape the assaults and still be able to care for my child.

My friend knew of no place I could go, nowhere I could turn, no one else I could ask. I remember saying in a tone of resignation, "I can't apply for welfare." Instead of disagreeing with me, she nodded, acknowledging our

SOURCE: Rita Henley Jensen, "Welfare," *Ms. Magazine* 6 (July 1995), pp. 56–61. Reprinted by permission of *Ms. Magazine*, © 1995.

mutual belief that taking beatings was better than taking handouts. Being "on the dole" meant you deserved only contempt.

In August 1965, I married my attacker.

Six years later, I left him and applied for assistance. My children were 18 months and five and a half years old. I had waited much too long. Within a year, I crossed High Street to go to Ohio State. I graduated in four years and moved to New York City to attend Columbia University's Graduate School of Journalism. I have worked as a journalist for 18 years now. My life on welfare was very hard—there were times when I didn't have enough food for the three of us. But I was able to get an education while on welfare. It is hardly likely that a woman on AFDC today would be allowed to do what I did, to go to school and develop the kind of skills that enabled me to make a better life for myself and my children.

This past summer, I attended a conference in Chicago on feminist legal theory. During the presentation of a paper related to gender and property rights, the speaker mentioned as an aside that when one says "welfare mother" the listener hears "black welfare mother." A discussion ensued about the underlying racism until someone declared that the solution was easy: all that had to be done was have the women in the room bring to the attention of the media the fact that white women make up the largest percentage of welfare recipients. At this point, I stood, took a deep breath, stepped out of my professional guise, and informed the crowd that I was a former welfare mother. Looking at my white hair, blue eyes, and freckled Irish skin, some laughed; others gasped—despite having just acknowledged that someone like me was, in fact, a "typical" welfare mother.

Occasionally I do this. Speak up. Identify myself as one of "them." I do so reluctantly because welfare mothers are a lightning rod for race hatred, class prejudice, and misogyny. Yet I am aware that as long as welfare is viewed as an African American woman's issue, instead of a woman's issue—whether that woman be white, African American, Asian, Latina, or Native American—those in power can continue to exploit our country's racism to weaken and even eliminate public support for the programs that help low-income mothers and their children.

I didn't have the guts to stand up during a 1974 reception for Ohio state legislators. The party's hostess was a leader of the Columbus chapter of the National Organization for Women and she had opened up her suburban home so that representatives of many of the state's progressive organizations could lobby in an informal setting for an increase in the state's welfare allotment for families. I was invited as a representative of the campus area's single mothers' support. In the living room, I came across a state senator in a just-slightly-too-warm-and-friendly state induced by the potent combination of free booze and a crowd of women. He quickly decided I looked like a good person to amuse with one of his favorite jokes. "You want to know how a welfare mother can prevent getting pregnant?" he asked, giggling. "She can just take two aspirin—and put them between her knees," he roared, as he bent down to place his Scotch glass between his own, by way of demonstration. I drifted away.

I finally did gather up my courage to speak out. It was in a classroom during my junior year. I was enrolled in a course on the economics of public policy because I wanted to understand why the state of Ohio thought it desirable to provide me and my two kids with only $204 per month—59 percent of what even the state itself said a family of three needed to live.

For my required oral presentation, I chose "Aid to Families with Dependent Children." I cited the fact that approximately two thirds of all the poor families in the country were white; I noted that most welfare families consisted of one parent and two children. As an audio-visual aid, I brought my own two kids along. My voice quavered a bit as I delivered my intro: I stood with my arms around my children and said, "We are a typical AFDC family."

My classmates had not one question when I finished. I don't believe anyone even bothered to ask the kids' names or ages.

If I were giving this talk today, I would hold up a picture of us back then and say we still represent typical welfare recipients. The statistics I would cite to back up that statement have been refined since the 1970s and now include "Hispanic" as a category. In 1992, 38.9 percent of all welfare mothers were white, 37.2 percent were black, 17.8 percent were "Hispanic," 2.8 percent were Asian, and 1.4 percent were Native American.

My report, however, would focus on the dramatic and unrelenting reduction in resources available to low-income mothers in the last two decades.

Fact: In 1970, the average monthly benefit for a family of three was $178. Not much, but consider that as a result of inflation, that $178 would be approximately $680 today. And then consider that the average monthly payment today is only about $414. That's the way it's been for more than two decades: the cost of living goes up (by the states' own accounting, the cost of rent, food, and utilities for a family of three has doubled), but the real value of welfare payments keeps going down.

Fact: The 1968 Work Incentive Program (the government called it WIN; we called it WIP) required that all unemployed adult recipients sign up for job training or employment once their children turned six. The age has now been lowered to three, and states may go as low as age one. What that means is you won't be able to attend and finish college while on welfare. (In most states a college education isn't considered job training, even though experts claim most of us will need college degrees to compete in the workplace of the twenty-first century.)

Fact: Forty-two percent of welfare recipients will be on welfare less than two years during their entire lifetime, and an additional 33 percent will spend between two and eight years on welfare. The statistics haven't changed much over the years: women still use welfare to support their families when their children are small.

In 1974, I ended my talk with this joke: A welfare mother went into the drugstore and bought a can of deodorant. I explained that it was funny because everyone knew that welfare mothers could not afford "extras" like personal hygiene products. My joke today would be: A welfare mother believed that if elected public officials understood these facts, they would not campaign to cut her family's benefits.

The idea that government representatives care about welfare mothers is as ridiculous to me now as the idea back then that I would waste my limited funds on deodorant. It is much clearer to me today what the basic functions of welfare public policy are at this moment in U.S. history.

By making war on welfare recipients, political leaders can turn the public's attention away from the government's redistribution of wealth to the wealthy. Recent studies show that the United States has become the most economically stratified of industrial nations. In fact, Federal Reserve figures reveal that the richest 1 percent of American households—each with a minimum net worth of $2.3 million—control nearly 40 percent of the wealth, while in Britain, the richest 1 percent of the population controls about 18 percent of the wealth. In the mid-1970s, both countries were on a par: the richest

1 percent controlled 20 percent of the wealth. President Reagan was the master of this verbal shell game. He told stories of welfare queens and then presided over the looting of the nation's savings and loans by wealthy white men.

Without a doubt, the current urgency for tax cuts and spending reductions can be explained by the fact that President Clinton tried to shift the balance slightly in 1992 and the wealthy ended up paying 16 percent more in taxes the following year, by one estimate.

The purpose of this antiwelfare oratory and the campaigns against sex education, abortion rights, and aid to teenage mothers is to ensure a constant supply of young women as desperate and ashamed as I was. Young women willing to take a job at any wage rate, willing to tolerate the most abusive relationships with men, and unable to enter the gates leading to higher education.

To accomplish their goals, political leaders continually call for reforms that include demands that welfare recipients work, that teenagers don't have sex, and that welfare mothers stop giving birth (but don't have abortions). Each "reform" addresses the nations' racial and sexual stereotypes: taking care of one's own children is not work; welfare mothers are unemployed, promiscuous, and poorly motivated; and unless the government holds their feet to the fire, these women will live on welfare for years, as will their children and their children's children.

This type of demagoguery has been common throughout our history. What sets the present era apart is the nearly across-the-board cooperation of the media. The national news magazines, the most prestigious daily newspapers, the highly regarded broadcast news outlets, as well as the supermarket tabloids and talk-radio hosts, have generally abandoned the notion that one of their missions is to sometimes comfort the afflicted and afflict the comfortable. Instead, they too often reprint politicians' statements unchallenged, provide charts comparing one party's recommendations to another's without really questioning those recommendations, and illustrate story after story, newscast after newscast, with a visual of an African American woman (because we all know they're the only ones on welfare) living in an urban housing project (because that's where all welfare recipients live) who has been on welfare for years.

When *U.S. News & World Report* did a major story on welfare reform this year, it featured large photographs of eight welfare recipients, seven of whom were women of color: six African Americans and one Latina or Native American (the text does not state her ethnicity). Describing the inability of welfare mothers to hold jobs (they are "hobbled not only by their lack of experience but also by their casual attitudes toward punctuality, dress, and coworkers") the article offers the "excuse" given by one mother for not taking a 3 P.M. to 11 P.M. shift: "I wouldn't get to see my kids," she told the reporter. You can't win for losing—should she take that 3-to-11 job and her unsupervised kids get in trouble, you can be sure some conservative would happily leap on her as an example of one of those poor women who are bad mothers and whose kids should be in orphanages.

Why don't the media ever find a white woman from Ohio or Iowa or Wisconsin, a victim of domestic violence, leaving the father of her two children to make a new start? Or a Latina mother like the one living in my current neighborhood, who has one child and does not make enough as a home health care attendant to pay for her family's health insurance? Or a Native American woman living on a reservation, creating crafts for pennies that will be sold by others for dollars?

Besides reinforcing stereotypes about the personal failings of welfare recipients, when my colleagues write in-depth pieces about life on welfare they invariably concentrate on describing welfare mothers' difficulties with the world at large: addictions, lack of transportation, dangerous neighbors, and, most recently, shiftless boyfriends who begin beating them when they do get jobs—as if this phenomenon were limited to relationships between couples with low incomes.

I wonder why no journalist I have stumbled across, no matter how well meaning, has communicated what I believe is the central reality of most women's lives on welfare: they believe all the stereotypes too and they are ashamed of being on welfare. They eat, breathe, sleep, and clothe themselves with shame.

Most reporting on welfare never penetrates the surface, and the nature of the relationship between the welfare system and the woman receiving help is never explored. Like me, many women fleeing physical abuse must make the welfare department their first stop after seeking an order of protection. Studies are scarce, but some recent ones of women in welfare-to-work programs across the U.S. estimate that anywhere from half to three fourths of participants are, or have been, in abusive relationships. And surveys of some homeless shelters indicate that half of the women living in them are on the run from a violent mate.

But if welfare is the means of escape, it is also the institutionalization of the dynamic of battering. My husband was the source of my and my children's daily bread and of daily physical and psychological attacks. On welfare, I was free from the beatings, but the assaults on my self-esteem were still frequent and powerful, mimicking the behavior of a typical batterer.

As he pounds away, threatening to kill the woman and children he claims to love, the abuser often accuses his victims of lying, laziness, and infidelity. Many times, he threatens to snatch the children away from their mother in order to protect them from her supposed incompetence, her laziness, dishonesty, and sexual escapades.

On welfare, just as with my husband, I had to prove every statement was not a lie. Everything had to be documented: how many children I had, how much I paid for rent, fuel, transportation, electricity, child care, and so forth. It went so far as to require that at every "redetermination of need" interview (every six months), I had to produce the originals of my children's birth certificates, which were duly photocopied over and over again. Since birth certificates do not change, the procedure was a subtle and constant reminder that nothing I said was accepted as truth. Ever.

But this is a petty example. The more significant one was the suspicion that my attendance at Ohio State University was probably a crime. Throughout my college years, I regularly reported that I was attending OSU. Since the WIN limit at that time was age six and my youngest daughter was two when I started, I was allowed to finish my undergraduate years without having to report to some job-training program that would have prepared me for a minimum-wage job. However, my caseworker and I shared an intuitive belief that something just had to be wrong about this. How could I be living on welfare and going to college? Outrageous! Each day I awoke feeling as if I were in a race, that I had to complete my degree before I was charged with a felony.

As a matter of fact, I remember hearing, a short time after I graduated, that a group of welfare mothers attending college in Ohio were charged with food stamp fraud, apparently for not reporting their scholarships as additional income.

Batterers frequently lie to their victims—it's a power thing. Caseworkers do too. For example, when I moved to

New York to attend graduate school and applied for assistance, I asked my intake worker whether I could apply for emergency food stamps. She told me there was no emergency food program. The kids and I scraped by, but that statement was false. I was unaware of it until welfare rights advocates successfully sued the agency for denying applicants emergency food assistance. In another case, when someone gave me a ten-year-old Opel so I could keep my first (very low paying) reporting job, my caseworker informed me in writing that mere possession of a car made me ineligible for welfare. (I appealed and won. The caseworker was apparently confused by the fact that although I was not allowed to have any assets, I did need the car to get to work. She also assumed a used car had to have some value. Not this one.)

Then there's the issue of sexual possessiveness: states rarely grant assistance to families with fathers still in the home. And as for feeling threatened about losing custody, throughout the time I was on welfare, I knew that if I stumbled at all, my children could be taken away from me. It is widely understood that any neighbor can call the authorities about a welfare mother, making a charge of neglect, and that mother, since she is less than nothing, might not be able to prove her competency. I had a close call once. I had been hospitalized for ten days and a friend took care of my children. After my return home, however, I was still weak. I would doze off on the sofa while the kids were awake—one time it happened when they were outside playing on the sidewalk. A neighbor, seeing them there unattended, immediately called the child welfare agency, which sent someone out to question me and to look inside my refrigerator to see if I had any food. Luckily, that day I did.

Ultimately, leaving an abusive relationship and applying for welfare is a little like leaving solitary confinement to become part of a prison's general population. It's better, but you are still incarcerated.

None of this is ever discussed in the context of welfare reform. The idiot state legislator, the prosecutor in Ohio who brought charges against welfare mothers years ago, Bill Clinton, and Newt Gingrich all continue to play the race and sex card by hollering for welfare reform. They continue to exploit and feed the public's ignorance about and antipathy toward welfare mothers to propel their own careers. Sadly, journalists permit them to do so, perhaps for the same reason.

Lost in all this are the lives of thousands of women impoverished by virtue of their willingness to assume the responsibility of raising their children. An ex-boyfriend used to say that observing my struggle was a little like watching someone standing in a room, with arms upraised to prevent the ceiling from pressing in on her. He wondered just how long I could prevent the collapse.

Today, welfare mothers have even less opportunity than I did. Their talent, brains, luck and resourcefulness are ignored. Each new rule, regulation, and reform makes it even more unlikely that they can use the time they are on welfare to do as I did: cross the High Streets in their cities and towns, and realize their ambitions. Each new rule makes it more likely that they will only be able to train for a minimum-wage job that will never allow them to support their families.

So no, I don't think all we have to do is get the facts to the media. I think we have to raise hell any way we can.

Our goal is simple: never again should there be a young woman, standing in front of the gates that lead to a better future, afraid to enter because she believes she must instead choose poverty and battery.

Real Welfare Reform

Janis Johnston

Janis Johnston is a divorced single mother who has experienced the welfare system first-hand. She is now a Ph.D. candidate in sociology at Colorado State University. In this essay she addresses issues of welfare policy, the media's role in promoting negative stereotypes about those on welfare, and the plight of poor children. She argues for the structural sources of poverty and the necessity of a strong welfare system. What do you think of her ideas concerning welfare policy?

The question of what we as an affluent society should provide for those who cannot provide for themselves appears to have no easy answer. In a perfect world there would be no discussion about welfare. Everyone would have enough to eat and a safe place to live. Everyone would have clothes, shoes, and a warm coat. Everyone would have adequate health care. But this isn't a perfect world; there are millions of Americans who do not have these things. We do have something in place to help those in poverty, and that thing is known generically as welfare. Individually, people tend to support welfare programs, though often with a series of requirements attached to them. Collectively, however, society frequently attacks the welfare system, and while millions of people benefit from the welfare system, it is interesting that for the most part the general public doesn't understand what welfare really is.

The bulk of welfare spending is on social insurance programs: Social Security, Medicare, nonmilitary retirement and disability benefits, unemployment insurance, and workers' compensation. The rest of welfare spending encompasses programs such as: Medicaid and medical aid for veterans; cash aid; food benefits, including food stamps, school lunch programs, and WIC; housing benefits; financial aid programs for education such as Pell Grants, Stafford Loans, and Head Start; followed by job and training programs, social services, and energy assistance. Of these programs, most public attention and indeed animosity is directed at only a small portion of welfare recipients: people who receive Aid to Families with Dependent Children (AFDC) or, since the welfare reform of 1996, Temporary Assistance to Needy Families (TANF). In fact, if you look at the media and listen to politicians, you may believe that these programs make up all the welfare spending in the United States.

When the press writes about welfare, they really mean TANF but they never call it TANF. Notice though, that when the same authors write about Social Security, they call it Social Security; they don't list it ambiguously under the title "welfare." When they write about subsidized loans for college students, they call them subsidized loans; they don't label them as "welfare." The popular conception of welfare, though, goes a long way to explain why the public feels the way it does about welfare. People hear about welfare in the narrow sense but associate it with the costs of welfare as a whole. The unstated but inferred idea is that all welfare spending goes to a select group. The reality, though, is that AFDC and TANF encompass only a small portion of the total social welfare budget and account for less than 1 percent of the total GNP in this country, which is usually far from what people believe. Aside from narrowly defining what welfare is, the press also provides the public with a composite of those who receive services, and that picture is overwhelmingly negative.

SOURCE: This essay was written expressly for this book and is published with the author's permission.

At a fundamental level, the question "Why does a program such as AFDC or TANF generate so much controversy?" is an important one, because it is the question that drives much of the welfare debate. When columnists and reporters write about welfare, it is almost always pejorative. When the columns are "positive," they reflect on how the changes in policy that require welfare recipients to go to work and on how those changes have reduced the number of people getting services. A recent issue of the *Rocky Mountain News* showed a woman riding a bus after work under the headline, "Welfare rolls down 56 percent." Of course it's not welfare rolls that are down, but rather the percentage of those receiving benefits under TANF. The press also uses other misleading tactics. I've seen headlines that read, "Welfare mothers should work too," which feed people's perceptions that welfare recipients don't work. In fact, if you listen only to the popular press, you may believe the typical welfare recipient is a young minority woman, raised on welfare herself, who had her first child when she was 15 and has never had a job. In reality, she is likely to be white and in her twenties, a first-generation welfare recipient with one or two kids and who has been on the system for less than five years.

The press is a primary reason that for most people the word "welfare" conjures up images of a single mother who has more and more babies just to get more and more money from the government, and all this because she is too lazy to work. Welfare policies are then attacked and changed based on this misperception, and the new policies are therefore doomed to fail because the stereotype is inaccurate. Thus, in order for real welfare reform to take place, the reforms should begin with the media. If reporters are talking about TANF, they should say TANF; if they mean welfare, they should include everything that falls under that umbrella. When they cry for welfare reform, they should either look at reforms for all of welfare or make the plea for reforms explicit to the program, usually TANF. Finally, people should understand what the reforms mean in terms of overall savings to the budget, and we should hold our politicians to the same standards. It is no longer acceptable to stereotype people because of race. Now let's add another category. Let's make it inappropriate to assume things about people just because they are poor. Let's try to understand normal people's stories rather than relying on the few exceptions that feed the myths.

Welfare reform often seems intent upon instilling a work ethic into those who qualify for benefits. Welfare recipients are not people who would rather get a handout than work for a living. Many of those on welfare have jobs of some kind, and many others have compelling reasons for not working. Many of those on welfare are simply too young to work. How, then, is forcing people off the system in two years or sending them unprepared into the working world supposed to help? You can't fix something unless you know what's broken, and people's desire to work is not the problem. Welfare reform as we know it focuses on the individual, but that is wrong. The focus should be on the structure. If the structure of society were more amenable to all individuals, we could greatly decrease the number of people who suffer the effects of poverty. The structure should provide safe daycare centers. The structure should provide access to education or job training programs, while at the same time understanding that two years may not be enough time to finish the training. Most critically, the poor need jobs that pay a living wage.

The structure should pay special attention to children. Children make up a large part of those who are deeply affected by poverty. In a country with all the wealth that this one has, it is unconscionable that anyone would go hungry, most especially children. Regardless of whether you believe that individuals are responsible for their poverty, children are clearly not responsible for their situations. They cannot pick their parents, and they cannot go out and support themselves. Poverty affects children in so many ways, especially at the most basic level. Let's consider something as simple as diet. Kids need proper nutrition, not only for their health now, but also for their health during their adult years. Providing an adequate diet for children when they are young will stave off health problems later in life. Therefore, from a purely economic perspective, kids should have milk, and bread, and peanut butter, and spaghetti, and even soda and chips; kids should have plenty to eat. Providing an adequate diet is advantageous beyond just providing a foundation for better health later in life; it gives them what they need to succeed in school. It's hard to focus on school when you're hungry.

Even if schools provide free or low-cost meals (which is still welfare), the benefits of a free lunch sometimes come at a high social price that kids aren't willing to pay. If kids are separated into "free lunch" kids and "paying" kids, everyone is reminded who the poor kids are on a daily basis. Even at young ages that difference stigmatizes. There are ways to calculate how many people will eat free lunches versus those who pay in a more traditional way without embarrassing kids or making them articulate their poverty over and over. We should ensure, through breakfast and lunch programs, that all children have enough to eat. In addition, the meals should be provided in ways that protect the privacy of the children.

School lunches are not the only food aid to the poor. Many of those who live in poverty rely on food stamps, and using food stamps is an interesting experience. There are rules that must be followed. For example, you can't buy anything that is maintained at "above room temperature." If you buy cold, ready-to-eat chicken that's ok, but you can't buy the same thing if it's sitting under a heat lamp. You also can't buy other items that are as important as food, like toilet paper, toothpaste, and soap. Beyond the rules, however, there are other aspects to shopping with food stamps that make the experience difficult.

By their very design, food stamps cannot be used without most people knowing that you are on government assistance. Because food stamps make you so visible, people often make comments about your purchases. I had people tell me that ice cream wasn't very nutritional and therefore they wanted to know why "their" tax dollars should pay for my junk food. It doesn't have to be that way. We could set up a system whereby people on food stamps would receive a debit card. The card could be credited each month with a dollar amount equivalent to what would normally be sent through the mail. This would save the government money and time in printing and mailing costs as well as giving the recipients some level of privacy. We could also change the food stamp system so that people could purchase the things they need. Is it really important to make sure that people on food stamps don't buy hot food or personal hygiene products?

Access to medical care is crucial, especially for those who are poor. Poor people have more health problems for a variety of reasons, but most often because they lack access to affordable care. I once knew a dentist who was very proud of his profession because dentists, above all others, work to put themselves out of business. At some level he is right. Dentistry has come a long way, and now some people can keep strong, healthy, real teeth for their whole lives. But not all people. People who can afford regular checkups and fluoride treatments and cleanings can keep strong, healthy, real teeth for their whole lives, but unless you have the money to pay up front, many dentists won't even allow you to make an appointment. I understand that dentists have bills to pay and demands on their cash for things like equipment, buildings, insurance, and assistants and therefore need to earn enough to afford all those things. But in the town I just moved from, there is not one dentist who will see a patient on Medicaid. Not one. And I think that is wrong.

In a perfect world everyone would have access to medical care, including preventive medicine. The lack of it is not only devastating to those who are being ignored, but it is an embarrassment to health-care providers and to all members of this society. Preventive care is also cheaper than treating emergencies. Again, however, it comes down to the structure of society. We have it within our power to provide health care to people who need it, but the structure does not encourage such behavior. In fact, the opposite behavior is rewarded. That needs to change.

In welfare, when one differentiates between "able-bodied" and non-able-bodied," the distinction necessarily carries with it ideas of "deserving" and "undeserving" poor. Those ideas are carried over into how we, as a society, treat welfare recipients. But it is interesting how we define deserving poor. Most people would agree that someone who is too ill to work should be provided with the basics, including help with housing, food, and medical care. Can we tell who fits into that category just by looking? What if the person is a young man who has a heart condition? Can we tell that he deserves assistance just by looking at him? How about the young woman who brings six children under the age of seven with her to the store and then uses food stamps to buy food for dinner. Does she deserve assistance? Does it matter if only one of the children is hers and she baby-sits for the rest of them to supplement her income? Should it matter?

People on welfare are people like me, and if you have been lucky enough not to have had welfare experience, I hope you never will. But if you ever do end up in such a situation, I hope help is there if you need it.

7

The Health Care System

There are two government health programs—Medicare for the elderly and Medicaid for the very poor. Virtually all of the elderly have at least some health coverage through these programs. The same is not true for the working poor and the near-poor, who are left out. The latter make up the one-third of the poor who are not covered by health insurance. A partial list of the problems with Medicaid includes the following. First, the programs are funded jointly by federal and state governments. Thus, coverage varies from state to state in quality, eligibility of patients, coverage, and the adequacy of fees paid to physicians and hospitals for their services. Second, many physicians refuse to treat Medicaid patients because the physicians are not reimbursed enough, which results in delayed treatment and treatment in public hospital emergency rooms where indigent patients cannot be turned away. This situation overburdens hospitals, resulting in long waits and hurried attention by overwhelmed health practitioners. Third, disproportionately fewer physicians practice in poor urban and poor rural areas than in affluent urban and suburban areas.

The poor, in contrast to the privileged, are less healthy. For instance, more than 20 percent of individuals in households with an income of less than $15,000 report being in fair or poor health, compared with about 4 percent of individuals in households with incomes of $50,000 or more (Henwood, 2000). The reasons are that the poor live in home, neighborhood, and work environments that are more stressful, less secure, and more dangerous. They are more likely to be exposed to toxic wastes and pollution in their neighborhoods. Affordable housing for the poor and the near-poor is often found where the air, water, and land are polluted. Exposure to lead that is typically found in the paint and in the plumbing in old buildings is especially dangerous for young children. Such exposure is related to behavioral problems, focus, reduced intelligence, and speech problems. Exposure to PCBs (polychlorinated biphenyls) and mercury—the byproducts of chemical and industrial production—cause birth defects, lack of coordination, diminished intelligence, and poor memory in children (Kaplan and Morris, 2000).

The impoverished have more stress and hypertension (high blood pressure) because they live in crowded conditions and worry about having enough resources for food, utilities, and rent. The poor also may not have adequate heat, ventilation, and sanitation, which means that they are more susceptible to infectious and parasitic diseases. The children of the poor have less healthy environments than more privileged children in the crucial first five years of life. They are more likely to have less healthy diets, less likely to be breastfed, more likely to be exposed to lead and other chemicals that negatively affect health and cognitive ability, and less likely to receive preventive care and early treatment.

Consequently, economic disadvantages are closely associated with health disadvantages. The poor are more likely than the affluent to suffer from certain forms of cancers (lung, cervix, and esophagus), hypertension, low birth weight, hearing loss, diabetes, and infectious diseases (especially influenza and tuberculosis). And, when stricken with disease, they are less likely to survive than the more well-to-do. These differences in health by family income are seen most dramatically for children in long-term poor families, who are, when compared with children from families at least four times above the poverty line:

- 1.5 to 3 times more likely to die in childhood.
- 2.7 times more likely to have stunted growth.
- 3 to 4 times more like to have iron deficiency as preschoolers.
- 1.5 to 2 times more likely to be partly or completely deaf.
- 1.2 to 1.8 times more likely to be partly or completely blind.
- about 2 times more likely to have serious physical or mental disabilities.
- 2 to 3 times more likely to have fatal accidental injuries.
- 1.6 times more likely to have pneumonia (Sherman, 1997:4).

Clearly, then, the delivery of health care is not evenly distributed in the United States, with dire consequences for the poor. The following selections highlight the difficulties of the impoverished as they attempt to deal with health problems with little or no help from society.

NOTES AND SUGGESTIONS FOR FURTHER READING

Children's Defense Fund (2001). *The State of America's Children: 2000.* Washington, DC: *Consumer Reports* (2000). "Second-Class Medicine." Volume 65 (September):42–50.

Digby, Anne (1996). *Gender, Health, and Welfare.* New York: Routledge.

Dollars and Sense (2001). "Putting Names on the Numbers: Testimonies of the Uninsured." Number 235 (May/June):32–33.

Duncan, Greg J., and Jeanne Brooks-Gunn (1997). *Consequences of Growing Up Poor.* New York: Russell Sage.

Federal Interagency Forum on Child and Family Statistics (2000). *America's Children: Key National Indicators of Well-Being 2000.* Washington, DC: U.S. Department of Health and Human Services.

Henwood, Doug (2000). "Health & Wealth." *The Nation* (July 10):10.

Hofrichter, Richard (ed.) (1993). *Toxic Struggles: The Theory and Practice of Environmental Justice.* Philadelphia: New Society Publishers.

Jones, Jacqueline (1992). *The Dispossessed: America's Underclasses from the Civil War to the Present.* New York: Basic Books.

Kaplan, Sheila, and Jim Morris (2000). "Kids at Risk." *U.S. News & World Report* (June 19):47–53.

Lefkowitz, Bonnie (2000). "Dollars Count More Than Doctors." <http://inequality.org/healthdc2.html>

Matteo, Sherri M. (1993). *American Women in the Nineties: Today's Critical Issues.* Boston: Northeastern University Press.

Reuss, Alejandro (2001). "Cause of Death: Inequality." *Dollars and Sense,* No. 235 (May/June):10-12.

Scher, Abby (2001). "Access Denied: Immigrants and Health Care." *Dollars and Sense,* Number 235 (May/June):8.

Sherman, Arloc (1997). *Poverty Matters: The Cost of Child Poverty in America.* Washington, DC: Children's Defense Fund.

U.S. Department of Health and Human Services (1999). *Health, United States, 1999.* Hyattsville, Maryland: Centers for Disease Control and Prevention (DHHS Publication 99-1232).

Getting to Work

Kenny Fries

Kenny Fries is an author and poet. Here he focuses on the effects of politics and the health care system on his own life. As you read this essay, consider the following: Much of the individual/cultural perspective on poverty focuses on the individuals' exhibiting a present-time orientation and the lack of a work ethic. What happens when poverty is combined with physical or mental disability? How would the individual/cultural perspective address this issue? How would a structural theorist approach this essay? How is the Social Security system set up so that it provides disincentives for working? What is the "medical model" of disability, and how does that model affect people with disabilities?

It's been ten years now since the passage of the Americans with Disabilities Act (ADA). When signed by President George Bush, it was called "the most far-reaching civil rights legislation since the Civil Rights Act of 1964." Many of us who live with disabilities thought the ADA would create an array of employment opportunities since it requires employers to make "reasonable accommodations" to allow those with disabilities to work. Considering the economic boom that has left many businesses without the workers they need, people with disabilities should be joining the work force in record numbers.

But this has not proved to be the case. While the unemployment rate for the entire population hovers around 4 percent, 70 percent of working-age adults with disabilities are still unemployed, a figure that basically has not changed since the ADA's passage.

One reason is that Social Security provides some harsh disincentives for people with disabilities who want to enter the work force.

Twelve years ago, due to issues related to my congenital disability (I am missing bones in both legs), I was no longer able to perform routine daily tasks I used to take for granted. I couldn't do my own grocery shopping, take out the garbage, or do the dishes. Still can't. And since I couldn't make my living as an arts administrator anymore, I applied for Social Security.

Initially, Social Security provided me just enough money to pay rent and utility bills while I adapted to a new way of life. After a year on Social Security Disability Income (SSDI), I became eligible for Medicare. Since I also qualified for Supplemental Security Income (SSI), a means-tested program whose benefits vary from state to state, I also was eligible for Medicaid. The two health insurance programs paid all of my medical bills, and Medicaid reimbursed me for travel to doctors' appointments. I was able to get food stamps, which helped keep me fed. And eventually I qualified for five hours per week of state-supported "homemaking" assistance for the daily tasks I could no longer do on my own.

During the past five years, I have tried other ways to support myself. In 1996, I received a two-book contract for a prose memoir and an anthology of work by writers with disabilities. I began teaching in a low-residency graduate writing program, and I was able to instruct my students mostly through the mail. I could work at home and keep my own schedule, which I need to do because of my disability. But this work—writing and teaching part time—comes with neither health insurance nor retirement benefits.

When I signed my book contract, I designed, with the help of a disability advocate, a Plan to Achieve Self-Support (PASS), one of the work incentives that allows a person with a disability on SSI to train for or begin work without diminishing your SSI check. I could not be where I now am if not for this program and the help of my SSI caseworker and a few other employees of the Social Security Administration.

However, because I eventually made more money than I was allowed to under the eligibility rules for SSDI, I lost my Medicare coverage last year. Luckily, I've still got Medicaid, and the Massachusetts Medicaid program is one with good benefits. But the doctor who has seen me since I was an infant works in New York City, and therefore I have to pay him out of pocket, and I have to shell out for any X-rays or other tests I need when I visit him.

Soon, I will have to make a decision about whether I can afford to keep working. Once I earn a certain amount, I will no longer qualify for Medicaid. I will also become ineligible for the state-supported "homemaking" help I need.

Although it is clear that my disability-related expenses will increase as I age, I cannot go over the $2,000 resource limit. This is the uppermost amount the statutes allow a person to own in cash and other liquid resources (usually excluding ownership of a home and a car) and still be eligible for SSI and its auxiliary benefits, which include Medicaid.

So now I'm faced with a choice: Do I cut back on my work so I can still get SSI, or do I risk working without these benefits?

I am not the only disabled person who is in such a quandary. I know of many others. One husband and his wife annulled their marriage so she could get the assistance she needs after multiple sclerosis left her unable to care for herself.

Another woman had to apply for a waiver from the Secretary of Health in order to marry her partner. Otherwise, she would have lost the benefits that pay for her respirator, which just happens to keep her alive.

A woman with spina bifida inherited $35,000 from her grandparents, which she invested. But because her bank account is over the $2,000 resource limit, she has lost her Medicaid benefits, which paid her $2,000 disability-related monthly drug bill. This woman wants to work part-time to help pay her medical expenses, but if she does she will lose her disability status, forcing her to go see doctors who may not understand her particular medical history or current physical situation.

According to SSI regulations, this woman could decide to "spend down" her inheritance to again be eligible for SSI and Medicaid. That is, instead of investing the money that can help her care for herself as she ages, she would have to spend $33,000 on items that would not be considered "resources." Although I can think of many enjoyable ways to spend more than $33,000 in less than a month's time, none of them makes sense considering the extra money it takes to live as a person with a disability.

Benefits for the disabled are terribly outdated. According to Social Security rules, "sustained gainful employment" is $700 per month. Once you earn more than that for nine months (called a "trial work period"), you are ineligible, which is why I lost my SSDI back in 1996. But who considers earning $700 per month "sustained gainful employment"? In many cities, $700 a month will not even cover rent.

Or consider the regulations SSI uses to calculate monthly "income exclusions"—income you are allowed to keep before it reduces the amount of your SSI check. There is only a $20 general income exclusion, and the earned income exclusion is $65, plus one-half of any additional income earned each month. And these regulations are nothing in comparison to the more byzantine rules it would take more than the space of this article to describe.

These regulations put enormous hurdles in our way. Many disabled persons want to work and are fully capable of doing so. But it's gotten to the point where we just can't afford it. We can't go without our insurance benefits, and we can't afford to pay for them ourselves.

Last December, President Clinton signed the Work Incentives Improvement Act, which was designed to address just this problem. But the government is still, in effect, telling us that we have to stay poor and unemployed if we want the security of the health insurance and other assistance we need. The new law leaves it to each state to set its own varying thresholds for deciding how much income you can make and still keep Medicaid. And the new regulations do not modify the outdated resource limits. My choice—to work, or to keep receiving the assistance I need—remains essentially the same.

There are other reasons why persons with disabilities have not entered the work force in higher numbers.

One is in the way the ADA is written, using the intentionally ambiguous term "reasonable accommodations." This can mean many things to many employers. And enforcement of the law has been spotty, at best.

Another part of the answer lies surely in the stereotypes and fears about disabled people that still pervade our society. Can you think of a movie that addresses the day-to-day life of a paraplegic once rehabilitation is over? Those who live with disabilities are not part of public discourse. That's because people with disabilities are not thought of as having day-to-day lives. Once a person is disabled, what else is there to know other than that they are still disabled or that they've "recovered"?

A debilitating medical model underlies our society's view of the disabled.

That model focuses on the physical impairment, which needs to be overcome or cured. Instead of seeing the forces outside the body, outside the impairment, outside the self as essential to a disabled person's successful negotiation with an often hostile society (whether the barriers be financial, physical, or discriminatory), this view of disability—where cure and eradication of difference are paramount goals—puts the blame squarely on the individual when a physical impairment cannot be overcome.

Historian Paul K. Longmore points out that the medical model creates a class of persons with disabilities who are "confined within a segregated economic and social system and to a socioeconomic condition of childlike dependency."

Thanks to disability rights advocates and artists with disabilities, this model is under assault. And the ADA has created a legal framework to begin viewing disability in a way that shifts the focus from the impairment to how, as social scientist Victor Finkelstein says, "the nature of society . . . disables physically impaired people."

The assumptions behind Social Security for the disabled remain mired in the past, just as the benefits do. It's time to take a closer look at these assumptions, and it's time to make the benefits correspond to our needs. Otherwise, they will continue, in practice, to undermine the goals of the ADA and to sabotage the attempts of those of us with disabilities who want to include work as part of our lives.

If we were able to work as much as we could when we could, to save and invest the money we earn without risking our disability status, health insurance, and other benefits, not only would we be better off, but so would the rest of society.

On the Margins: The Lack of Resources and the Lack of Health Care

Janis Johnston

Adequate health care is something that many Americans take for granted, but health care can be a nightmare for those living in poverty. Janice Johnston is a divorced single mother and a graduate student in sociology at Colorado State University. In this essay, she reflects on what it is like to be poor and deal with a problematic health-care system. What are some of the problems that she encountered? Can you think of possible alternatives to this system?

Some days were harder than others. After 13 years of marriage I was in the middle of a divorce, had two kids—Lindsay who was 12 and James who was 6—and I was on unpaid disability leave from the job I'd had for ten years. My kids and I took our clothes and a box of dishes and moved forty-five miles to a town of about 25,000 so I could attend the university there. Because I couldn't work I thought I'd try school, but it was difficult because our only income was from the child support check we received every month.

We didn't have a car so we walked to most of the places that we needed to go and took the bus to school. Because we got only $300 a month, I supplemented our food by finding treasures that the nearest grocery store had thrown into

SOURCE: This essay was written expressly for this book and is published with the permission of the author.

the garbage: old bread, dented cans, and occasionally even frozen meat. It wasn't fun, but I had heard so many horror stories about welfare that I was afraid to go on it, afraid of how people—including members of my own family—would react. That lasted until the night the police came and got me out of a school concert because Lindsay was in the emergency room. While she was riding her bike home from the neighborhood convenience store, a car traveling about 30 miles an hour ran through a crosswalk and hit her, sending her skidding down the pavement. She was unconscious when the ambulance arrived but woke up as they worked on her. She immediately began to protest, saying we didn't have enough money for her to go in an ambulance and we surely couldn't afford any hospital. Because of her age they took her anyway.

I was devastated. My daughter was hurt and in pain, and rather than screaming for help she was trying to run away from it because we couldn't afford it. As much as that hurt, the reality was that she was right. We couldn't afford it. Her three-block ride to the hospital alone cost more than we lived on for a month. The cost of the emergency room, the doctor, and the x-rays would cost more than we received in a year. To add insult to injury, because she was riding a bike, it was determined that the accident was her fault although she was in a crosswalk and three lanes of traffic had stopped to allow her to cross; the bills were ours alone. The driver of the car called to say his insurance wouldn't pay our bills because it wasn't his fault, and he wondered what I intended to do about the damage to his car. I applied for welfare the next day.

On my walk to the social services center where the local welfare office was located, I kept telling myself over and over that it was OK—that applying for welfare didn't mean all those things that I'd heard people say it meant. Applying for welfare didn't mean I was lazy; that I wanted a handout for doing nothing; that I deserved what was happening, and worse, so did my kids. Welfare wasn't who I was at all and I wasn't going to let it define me. After all, I'd been working since I was 14, and for most of my married life I'd held two or three jobs at a time. I wasn't looking for a freebie, but I couldn't make it on my own until I could go back to work. If I could just get a little help—just a little—we could make it. By the time I got to the office, I thought I had myself fairly convinced. I went into the office, filled out the required paperwork, and made an appointment with a caseworker. I stepped out of the building and thought, "There, that wasn't so bad." And then I cried like I would never stop.

We were lucky. My application for welfare was dated the last day of October, which meant that if it were approved, the medical care portion, Title XIX, would cover any medical costs that occurred that month. Lindsay's accident was on October eleventh, so if we met the criteria, all of those hospital bills would be paid, follow-up care would be available, and we would get food stamps to help with the cost of food. All I had to do now was get over the idea that being on welfare exposed some inherent flaw in my character. I might actually have been able to do just that if so many people hadn't been telling me welfare really *was* me.

My visit with my caseworker was my introduction to the way people feel about welfare moms. We went over my application to make sure I understood and answered all the questions. Some of the questions seemed ludicrous. One section asked if "anyone in the house owned or was buying any of the following" and then listed 29 possible assets including: "Antiques/Collectibles; Airplane; Business Inventory; and Burial Policy." Airplane?! It was all I could do not to ask the caseworker whether my

Lear jet would disqualify me from receiving the $360 a month stipend the state paid for families of three; it wasn't a plane after all. Luckily I didn't mention it because this was a woman who found no joy in her job. Although she never said it directly, it was clear that she didn't understand why I was there. During the interview she mumbled that *she* had two kids and *she* had put herself through college and *she* never needed assistance from the government. I didn't need her to finish her thought; what she'd said was enough to let me know what she thought I deserved. Her treatment of me was a taste of what I would find when I went out into the "real world."

Life on welfare was tough for my kids. I mean we do things as a society to make sure it's obvious who *those people* are, who I am, who Lindsay and James are. It was common for the teacher to say, "Raise your hand if you have a free lunch ticket" to separate the kids who paid for their lunches from those who didn't. In supermarkets people with food stamps had to tell the clerk it was a food stamp order before they rang it up. Announcements such as "Food stamp change to checkstand three." were made on the loud speaker. Whether by design or by accident, all these actions keep welfare recipients in the public eye. For me this meant people could watch what I did, comment on what I bought, and question the way I dressed.

I was grocery shopping one day and a stranger noticed the food stamps in my hand as we waited at the register. I became aware that he was reviewing everything in my cart as well as noting what I was wearing. After a few minutes, he looked right at me and said, "I suppose you think my tax dollars are supposed to buy ice cream and chocolate syrup. And those sandals you have on look pretty expensive. I suppose I bought those for you too." He then backed his cart out of the line and went to the next closest one where he continued to stare at me until my transaction was complete and I walked out the door. Lesson number one. People on welfare have two choices. First, learn how to dress to shop with food stamps —not too nice, not too shabby—and buy only food that's good for you. Second, grow a thick skin to insulate yourself from the stares and comments. I did a little of both.

One of the things about being on welfare that I never counted on was what to do about medical care. Despite the fact that I was on disability from my job, my eligibility for the short-term disability insurance was only good for six months so it was gone. I wasn't eligible for medical care through the university because I wasn't a full-time student, and that meant that if I needed to see a doctor, or if my kids needed to see a doctor, we would have to go as Title XIX patients. Honestly, the pediatricians who took care of the kids were great. It was kind of a pain, because every time we went to see one of them we had to take a coupon, and if we didn't have one—if it hadn't arrived yet or was lost—the kids couldn't be seen. But it was a whole different story when it came to dentists, eye doctors, and doctors for me—worse if we were out of town.

A couple months after we went on welfare, I decided that we needed to get out of town for a break. We still didn't have a car, but a friend let me borrow hers so I could take my kids to see their dad. James had had a cold that I thought was getting better, but on the 45-mile ride to see my ex-husband, it was obvious to me that he was still coughing an awful lot and his fever was back. When we arrived in town, it was a little after 7:00 P.M., so all the doctors' offices were closed. I took him to a "doc-in-a-box," one of those emergency clinics you can go to without an appointment and asked how much it would cost for him to be seen by a doctor. The going rate for a kid with a bad cold was $30.00, plus medicine of course. According to the receptionist, the clinic didn't take Title XIX and they didn't take charity cases so we left to pool our pennies.

My ex-husband wasn't home—we hadn't called because we didn't have a phone—so the kids couldn't see him. Even worse, I couldn't get any money from him to take James to the doctor. I drove to the place where I'd worked until that August. My former co-workers scratched together $27 so I put that with $2.27 that I had and went back to the clinic. I was sure that $29.27 was close enough, especially when you could see how really sick this poor little boy was, but when we went in the office staff sent us away. They rationalized that the office visit was $30.00, not $29.27, and even if the doctor did see us it wouldn't matter because then we wouldn't have anything left to pay for medicine.

I was furious and at the same time worried and incredibly depressed. Here was a little boy who was getting sicker by the minute—I was starting to think it might be pneumonia—and there wasn't a damn thing I could do about it. We left in spite of my overwhelming desire to throw a big enough fit that they would have to call the cops. I mean at some level I thought if the cops were involved they would see the injustice and make them treat my son, but I didn't have the energy to take the chance and, after all, the cops would probably see things the way the people at the doctor's office did. I always felt that way—like I didn't deserve the same level of respect "normal" people got. But I hated the fact that my kids were being treated that way when there wasn't a fucking thing they could do about it! I was the one who was not working! I was the one who was on welfare! There was nothing about this situation that was their fault and nothing about the way people treated us that they deserved. We drove on.

There was another doc-in-a-box just a few miles from the first one, and we went there hoping that we had enough money for them. This time we had an entirely different experience. The office staff took one look at my son and took us into the inner office immediately. The doctor diagnosed him with an ear infection, bronchitis, and he was dangerously close to pneumonia. They gave us samples of cough medicine and children's Tylenol, enough to last for a couple weeks, and arranged for him to get the antibiotics. Then they refused our $29.27. I was overwhelmed. All the emotions, all the rage and fear and pain that I had carried from the first office seemed to melt and I was suddenly exhausted. I tried to thank them but they insisted it was nothing, headed us toward the car, and recommended that we get James home to bed. Just when I'd given up, there were people in the world who cared, and it wasn't based on how much money we had. I really needed to know that right then.

We weren't quite that lucky when it came to dentists. I tried to find a new dentist when we moved, and the receptionist at the first office I called asked what kind of insurance we had. When I told her we were on Title XIX she explained that the office already had all of *those* people they were required to take and that we would have to look elsewhere. So did the receptionist at the next office, and the next office, and in fact, so did the receptionists at the next seven offices. At the eighth office I was told that they didn't have any more room for people on Title XIX but that if it was an emergency I could go to the emergency room. Going to the emergency room to see if we had any cavities had never crossed my mind, so I asked if they had dentists there. The woman responded that "No, emergency rooms do not have dentists but if it is an emergency situation they can call a dentist who will then come and remove the offending tooth." "But none of these teeth are offensive," I responded. "We're just trying to keep them from getting that way." "Oh," she said, sounding almost disappointed, "I see. Well I don't know anyone who's taking Title XIX right now. You can call back in six months." I quit calling anywhere after that.

Eye appointments were about as easy to get as dental appointments. Lindsay started wearing glasses in the second grade. I wore glasses too, and I was getting headaches and things were blurry, so I thought I'd see if there was any hope that we could see an eye doctor. The short answer was no; Title XIX doesn't cover eyes. Then I got tricky and called to see if I could make an appointment and arrange a way to make payments. They asked if I had insurance and I said no. So did they. This was when family came in handy. For birthday presents my family would take us to the eye doctor and pay for the exam and even glasses if we needed them. I don't know what people on welfare do for dental or eye care if they aren't lucky enough to have access to other resources. I do know what happens to adults, however, when they need to go to a doctor and they're on welfare.

As I mentioned, I was on medical disability. Sometimes that meant that I was in a lot of pain, and sometimes it meant I had a hard time walking, but it was also episodic; I'd be OK for a while and then hit a bad period that might last a couple weeks. In the past, there were times when the doctors treated the injury with medicine alone, and sometimes it was a combination of medicine and therapy, but for the most part it was never an emergency. I knew that I needed to find a doctor who would help when it started to get bad and began looking for someone shortly after we moved. Time and time again I was told that the various offices were either not accepting new patients or that there was a significant wait to get in, which was OK as long as I was feeling pretty good. Then I hit a bad period. The doctors who didn't have room for me before had no room for me now so, as was the case with the dentist's office, it was suggested I seek help at the emergency room. This time I took them up on it.

I went into the hospital and asked to see a physician, filled out the paperwork, and waited. When the doctor came to see what my complaint was, I explained it and told him the kind of treatment I usually received—often a combination of painkillers, muscle relaxants, and anti-inflammatory medicines. He looked at me for a second, obviously thinking about something, and then asked why I had come to the emergency room because this was obviously not an emergency. I explained that I was having trouble getting in to see a "regular" doctor because many of them were not accepting new patients, and to get in to see those who were would take several months to a year. So while my case wasn't a real emergency, I couldn't wait that long to be seen either. He was clearly upset because there were real emergencies in the hospital and I was taking time away from them. And although he didn't say anything to me and was probably angry at the overall situation, I felt as if it was my fault. It was my fault that there were people who needed treatment that was being delayed because someone who could easily wait was taking up valuable space and time. The situation was awkward because although I needed help, I didn't need it from them, but I couldn't get it anywhere else. They treated me between the real cases and several hours later I went home.

I suppose it wouldn't have been a bad experience if it had happened only once, but the truth is, I began to treat the ER like a doctor's office. They didn't give me excuses as to why they didn't have to take any more welfare cases or why they weren't taking new patients. They didn't ask me to wait for months before I saw a doctor. But I could see the levels of frustration increase. I overheard two of the ER workers talking about people who used their services when it was obviously not an emergency. One of them mentioned a patient who had been in several times in just a couple weeks, and they couldn't get through to him that he needed to find a doctor and quit wasting valuable time and resources in the emergency room. And while I knew they weren't talking about me, they might as well have been. I struggled to find other options but it always came down to the same thing: Go to the emergency room or live with the debilitating pain and risk further damage.

See, that's what most people don't think about with living in poverty. Simple things. Routine exams. Getting your teeth cleaned. Finding a cavity and doing something about it before it means a root canal. Back when I had a real job, I had had a root canal on a back molar and then a crown. Years later when I was on welfare, I lost the crown and was told Title XIX wouldn't cover a replacement. Eventually it got to the point where something had to be done about the tooth and I returned to the dentist who had done the original work. I had known this man for years and even considered a friend. He saw the problem and was immediately angry with me. "How could you let this happen? You should have been in for another crown as soon as you lost the first one. This has never happened to *me* before. I've never had to pull a tooth that I saved." He pulled the tooth and left the room. His nurse finished the work and said she was surprised; she'd never known him to get mad like that, but he was really upset. I never went back.

That one trip to the dentist summed up my time on welfare as well as anything. My "friend" was upset at what I hadn't done because of how it affected *him,* how it made *him* look. And it is one way I will pay for being on welfare for the rest of my life. I have dental problems that I never would have had if I'd had access to preventative care—but I didn't. And I will live with that forever. Some people that I've known for years couldn't deal with my poverty so they are out of my life altogether now, and that too, will be forever. But we were lucky. It could have been worse. My kids and I might have had to exist that way for our lifetimes. For us, it only felt that long.

8

Schools and Schooling

When inadequate health care and inadequate diet are added to exposure to toxic chemicals, they reduce the odds of poor children succeeding in school. "At age 5, poor children are often less alert, less curious, and less effective at interacting with their peers than are more privileged youngsters" (Hewlett, 1991:56). Moreover, more privileged youngsters are more likely to attend enriched preschool programs than the children of the poor, again increasing the readiness gap between the privileged and the economically disadvantaged. When attending school, poor children are much more likely than privileged children to find their classrooms inadequately staffed, overcrowded, and ill equipped. Schools in U.S. society are financed primarily through local property taxes. This financing results in affluent districts having significantly greater resources for educating their children than economically marginal districts have for their children (Kozol, 1991). As a result, poor children are more likely than affluent children to attend schools that need repair, have fewer books and supplies, and have teachers with less training.

Children from poor families would have a better chance to succeed in school if they were exposed to enriched preschool experiences. Head Start is a program that provides positive results for children, but it is underfunded, and only 40 percent of those eligible receive its benefits. Meanwhile, more affluent children attend a variety of early development programs, have tutors to help them succeed, and receive educational advantages such as home computers, travel experiences, and visits to zoos, libraries, and various cultural activities. As a result, children of the impoverished tend to underachieve.

Through the voices of the poor the essays in this chapter show how the structure of education results in a disproportionate number of poor children failing in school.

NOTES AND SUGGESTIONS FOR FURTHER READING

Ayers, William, and Patricia Ford (eds.) (1996). *City Kids, City Teachers: Reports from the Front Row*. New York: The New Press.

Books, Sue (ed.) (1998). *Invisible Children in the Society and Its Schools*. Mahwah, NJ: Lawrence Erlbaum Associates.

Children's Defense Fund (2001). *The State of America's Children*. Washington, DC.

Hewlett, Sylvia Ann (1991). *When the Bough Breaks: The Cost of Neglecting Our Children*. New York: Basic Books.

Kozol, Jonathan (1991). *Savage Inequalities: Children in America's Schools*. New York: Crown.

Levin, Murray (1998). *Teach Me! Kids Will Learn When Oppression Is the Lesson*. New York: Monthly Review Press.

McNeil, Linda M. (2000). "Creating New Inequalities: Contradictions of Reform." *Phi Delta Kappan* 81 (June): 729–734.

Symonds, William C. (2001). "How to Fix America's Schools." *Business Week* (March 19):67–80, and the editorial on page 118.

Traub, James (2000). "What No School Can Do." *New York Times Magazine* (January 15) http://nytimes.com/library/magazine/home/20000116mag-traub8.html

U.S. Department of Education (1999). *Education Statistics Quarterly* 1 (Summer). Washington, DC: National Center for Education Statistics (NCES 1999-628).

Seventh Grade Disaster

Delphia Boykin

This essay was written by an African American woman about her experience as a volunteer in a Southside Chicago inner-city middle school. Analyze this reading using the sociological perspective regarding the impact of institutions on individual lives. From this selection, what do you predict about the future of these seventh grade students? How would you analyze the environment and life chances of these students from each theoretical perspective? What are some possible solutions to the high turnover rate of teachers in inner-city schools?

SOURCE: Delphia Boykin, "Seventh Grade Disaster," unpublished essay. Published with permission of the author.

During a conversation with a friend, she mentioned that her daughter brought home a note from school asking for volunteers to help with the students. I thought about applying, and it appealed to me because child development is a field that interests me for a career.

From September of 1998 through September, 1999, I volunteered to work in the Goldsmith Middle School. When I got there, the principal decided it would be best if I helped out a little in the seventh grade classroom. I was not prepared with any degrees, I had never worked with children before, and I had been out of school for ten years.

My first day there I thought, "This will be fun." I walked into the classroom and a nice elderly man named Mr. Nelson greeted me. I thought that we would be together for the rest of the year. Then he informed me that he was a substitute teacher. Mr. Nelson and I did well for a week, but then he left.

Within that school year, Mr. Jones, the superintendent, changed teachers five times. These frequent adjustments really affected the students. The children were so confused at this period that we had only four or five kids passing. The atmosphere was very chaotic.

Each teacher demonstrated a unique personality and method of teaching. The main subject in the seventh grade was math. One teacher named Mrs. Moody taught straight from the book. On the other hand, Mrs. Wright composed her lesson plans. She followed Mrs. Moody by only one month, so anyone can understand the confusion in the students.

The enrollment in the seventh grade was thirty-seven students crowded into a closet-size room approximately 8½ by 11 feet. This made it difficult to learn. The children were so close together that they could not bend their elbows. Under these circumstances, they were encouraged to play and talk. They would engage in conversations about their interests, with no regard for the teacher.

I can remember one day in the middle of winter, the snow was about six inches deep. I walked into the school and was told the teacher would not be in so I had to teach the class alone. The teacher had abruptly resigned. During that time I learned how difficult it is to work with many children at one time. There I was in a class with thirty-seven kids and no idea about how to manage them.

The next week another teacher came; this made me very happy, not for my sake, but for the sake of the students. I thought, "A new teacher—we can get down to business." Everything was going fine for three months and then he left.

The students had an equivalency test the week that he left, which made this the most challenging week of my tour of duty. I knew nothing about giving a test. I could barely follow the instructions for taking one myself. Naturally they told me the time limit for each test, but it was hard to monitor the kids and the clock. I could not wait for it to be over. I said to myself, "I made it through again."

One day in January I started gathering my papers to take them home to grade. Mr. Hobbs, the vice principal, came in and asked if I would mind doing the report cards. I said yes, even though I had to grade sixth, seventh, and eighth grade math papers. When the parents came to get the grades, they were very frustrated with the whole school system and there was nothing I could do about the situation.

I really felt alone because Mr. Ivy, the sixth grade teacher, and Mrs. Woodard, the eighth grade teacher, had fewer problems because of their past years of experience. Mrs. Woodard helped me a great deal, but it was impossible for her to teach my class and hers also.

When I started volunteering in the school, I thought it would be a great learning experience. Of course, we had training classes weekly for a month, but it was not adequate to prepare me for this horrid experience.

Mrs. Gray, the principal, had her hands full. Three schools were combined and she was the principal of all three schools. The years prior to my arrival had run much smoother because the previous seventh grade teacher had stayed for eight years before retiring. When the new teachers were hired, they wanted more money and better benefits. Mrs. Gray disagreed with this change so she continued to get substitutes.

Due to the circumstances, most of the children were failing. This was not their fault; they could not learn under these conditions. One student, Michael, wanted to excel but was failing because, as soon as he learned one method of teaching, another teacher arrived with a different method of teaching.

When I reflect upon this time, I realize there are many problems in the school system. The classrooms are crowded, the students are under taught, and the turnover of teachers is rapid.

Always Running

Luis Rodriguez

Luis Rodriguez immigrated from Mexico at a young age. This excerpt from his life story demonstrates the difficulties faced by children who are non-English speakers. Many states have adopted English-only laws that affect the school system. Would Luis' experiences have been different if he had been taught in his own language, or is total immersion in English the best method of teaching? What faults of the school system are highlighted in this excerpt? How do you think the different theoretical perspectives would approach the teaching of poor/minority students whose first language is not English?

Our first exposure in America stays with me like a foul odor. It seemed a strange world, most of it spiteful to us, spitting and stepping on us, coughing us up, us immigrants, as if we were phlegm stuck in the collective throat of this country. My father was mostly out of work. When he did have a job it was in construction, in factories such as Sinclair Paints or Standard Brands Dog Food, or pushing doorbells selling insurance, Bibles, or pots and pans. My mother found work cleaning homes or in the garment industry. She knew the corner markets were ripping her off but she could only speak with her hands and in choppy English.

Once my mother gathered up the children and we walked to Will Rogers Park. There were people everywhere. Mama looked around for a place we could rest. She spotted an empty spot on a park bench. But as soon as she sat down an American woman, with three kids of her own, came by.

"Hey, get out of there—that's our seat."

My mother understood but didn't know how to answer back in English. So she tried in Spanish.

"Look spic, you can't sit there!" the American woman yelled. "You don't belong here! Understand? This is not your country!"

Mama quietly got our things and walked away, but I knew frustration and anger bristled within her because she was unable to talk, and when she did, no one would listen.

We never stopped crossing borders. The Rio Grande (or *Rio Bravo,* which is what the Mexicans call it, giving the name a power "Rio Grande" just doesn't have) was only the first of countless barriers set in our path.

We kept jumping hurdles, kept breaking from the constraints, kept evad-

SOURCE: *Always Running—La Vida Loca, Gang Days in L.A.* by Luis J. Rodriguez. (Curbstone Press, 1993) Reprinted with permission of Curbstone Press. Distributed by Consortium.

ing the border guards of every new trek. It was a metaphor to fill our lives—that river, that first crossing, the mother of all crossings. The Los Angeles River, for example, became a new barrier, keeping the Mexicans in their neighborhoods over on the vast east side of the city for years, except for forays downtown. Schools provided other restrictions: don't speak Spanish, don't be Mexican—you don't belong. Railroad tracks divided us from communities where white people lived, such as South Gate and Lynwood across from Watts. We were invisible people in a city which thrived on glitter, big screens, and big names, but this glamour contained none of our names, none of our faces.

The refrain "this is not your country" echoed for a lifetime.

First day of school.

I was six years old, never having gone to kindergarten because Mama wanted me to wait until La Pata became old enough to enter school. Mama filled out some papers. A school monitor directed us to a classroom where Mama dropped me off and left to join some parents who gathered in the main hall.

The first day of school said a lot about my scholastic life to come. I was taken to a teacher who didn't know what to do with me. She complained about not having any room, about kids who didn't even speak the language. And how was she supposed to teach anything under these conditions! Although I didn't speak English, I understood a large part of what she was saying. I knew I wasn't wanted. She put me in an old creaky chair near the door. As soon as I could, I sneaked out to find my mother.

I found Rano's class with the mentally disabled children instead and decided to stay there for a while. Actually it was fun; they treated me like I was everyone's little brother. But the teacher finally told a student to take me to the main hall.

After some more paperwork, I was taken to another class. This time the teacher appeared nicer, but distracted. She got the word about my language problem.

"Okay, why don't you sit here in the back of the class," she said. "Play with some blocks until we figure out how to get you more involved."

It took her most of that year to figure this out. I just stayed in the back of the class, building blocks. It got so every morning I would put my lunch and coat away, and walk to my corner where I stayed the whole day long. It forced me to be more withdrawn. It got so bad, I didn't even tell anybody when I had to go to the bathroom. I did it in my pants. Soon I stunk back there in the corner and the rest of the kids screamed out a chorus of "P.U.!" resulting in my being sent to the office or back home.

In those days there was no way to integrate the non-English-speaking children. So they just made it a crime to speak anything but English. If a Spanish word sneaked out in the playground, kids were often sent to the office to get swatted or to get detention. Teachers complained that maybe the children were saying bad things about them. An assumption of guilt was enough to get one punished.

A day came when I finally built up the courage to tell the teacher I had to go to the bathroom. I didn't quite say all the words, but she got the message and promptly excused me so I didn't do it while I was trying to explain. I ran to the bathroom and peed and felt good about not having that wetness trickle down my pants leg. But suddenly several bells went on and off. I hesitantly stepped out of the bathroom and saw throngs of children leave their class. I had no idea what was happening. I went to my classroom and it stood empty. I looked into other classrooms and found nothing. Nobody. I didn't know what to do. I really thought everyone had gone home. I didn't bother to look at the playground where the whole school had been assembled for the fire drill. I just went home. It got to be a regular thing there for a while, me coming home early until I learned the ins and outs of school life.

Not speaking well makes for such embarrassing moments. I hardly asked questions. I just didn't want to be misunderstood. Many Spanish-speaking kids mangled things up; they would say things like "where the beer and cantaloupe roam" instead of "where the deer and antelope roam."

That's the way it was with me. I mixed up all the words. Screwed up all the songs.

"You can't be in a fire and not get burned."

This was my father's response when he heard of the trouble I was getting into at school. He was a philosopher. He didn't get angry or hit me. That he left to my mother. He had these lines, these cuts of wisdom, phrases and syllables, which swept through me, sometimes even making sense. I had to deal with him at that level, with my brains. I had to justify in words, with ideas, all my actions—no matter how insane. Most of the time I couldn't.

Mama was heat. Mama was turned-around leather belts and wailing choruses of Mary-Mother-of-Jesus. She was the penetrating emotion that came at you through her eyes, the mother-guilt, the one who birthed me, who suffered through the contractions and diaper changes and all my small hurts and fears. For her, dealing with school trouble or risking my life was nothing for discourse, nothing to debate. She went through all this hell and more to have me—I'd better do what she said!

Mama hated the *cholos.* They reminded her of the rowdies on the border who fought all the time, talked that *calo* slang, drank mescal, smoked marijuana, and left scores of women with babies bursting out of their bodies.

To see me become like them made her sick, made her cringe and cry and curse. Mama reminded us how she'd seen so much alcoholism, so much weed-madness, and she prohibited anything with alcohol in the house, even beer. I later learned this rage came from how Mama's father treated her siblings and her mother, how in drunken rages he'd hit her mom and drag her through the house by the hair.

The school informed my parents I had been wreaking havoc with a number of other young boys. I was to be part of a special class of troublemakers. We would be isolated from the rest of the school population and forced to pick up trash and clean graffiti during the rest of the school year.

"Mrs. Rodriguez, your son is too smart for this," the vice principal told Mama. "We think he's got a lot of potential. But his behavior is atrocious. There's no excuse. We're sad to inform you of our decision."

They also told her the next time I cut class or even made a feint toward trouble, I'd be expelled. After the phone call, my mom lay on her bed, shaking her head while sobbing in between bursts of how God had cursed her for some sin, how I was the devil incarnate, a plague, testing her in this brief tenure on earth.

My dad's solution was to keep me home after school. Grounded. Yeah, sure. I was thirteen years old already. Already tattooed. Already sexually involved. Already into drugs. In the middle of the night I snuck out through the window and worked my way to the Hills.

At sixteen years old, Rano turned out much better than me, much better than anyone could have envisioned during the time he was a foul-faced boy in Watts.

When we moved to South San Gabriel, a Mrs. Snelling took a liking to

Rano. The teacher helped him skip grades to make up for the times he was pushed back in those classes with the retarded children.

Mrs. Snelling saw talent in Rano, a spark of actor during the school's thespian activities. She even had him play the lead in a class play. He also showed some facility with music. And he was good in sports.

He picked up the bass guitar and played for a number of garage bands. He was getting trophies in track-and-field events, in gymnastic meets, and later in karate tournaments.

So when I was at Garvey, he was in high school being the good kid, the Mexican exception, the barrio success story—my supposed model. Soon he stopped being Rano or even José. One day he became Joe.

My brother and I were moving away from each other. Our tastes, our friends, our interests were miles apart. Yet there were a few outstanding incidents I fondly remember in relationship to my brother, incidents which despite their displays of closeness failed to breach the distance which would later lie between us.

When I was nine, for example, my brother was my protector. He took on all the big dudes, the bullies on corners, the ones who believed themselves better than us. Being a good fighter transformed him overnight. He was somebody who some feared, some looked up to. Then he developed skills for racing and high-jumping. This led to running track and he did well, dusting all the competition.

I didn't own any talents. I was lousy in sports. I couldn't catch baseballs or footballs. And I constantly tripped when I ran or jumped. When kids picked players for basketball games, I was the last one they chose. The one time I inadvertently hit a home run during a game at school—I didn't mean to do it—I ended up crying while running around the bases because I didn't know how else to react to the cheers, the excitement, directed at something I did. It just couldn't be me.

But Rano had enemies too. There were two Mexican kids who were jealous of him. They were his age, three years older than me. One was named Eddie Gambits, the other Rick Corral. One time they cornered me outside the school.

"You José's brother," Eddie said.

I didn't say anything.

"Wha's the matter? Can't talk?"

"Oh, he can talk all right," Ricky chimed in. "He acting the *pendejo* because his brother thinks he so bad. Well, he ain't shit. He can't even run."

"Yeah, José's just a *lambiche,* a kiss ass," Eddie responded. "They give him those ribbons and stuff because he cheats."

"That's not true," I finally answered. "My brother can beat anybody."

"Oh, you saying he can beat me," Eddie countered.

"Sure sounds like he said that," Ricky added.

"I'm only saying that when he wins those ribbons, *esta derecho,*" I said.

"It sounds to me like you saying he better than me," Eddie said.

"Is that what you saying, man?" Ricky demanded. "Com' on—is that what you saying?"

I turned around, and beneath my breath, mumbled something about how I didn't have time to argue with them. I shouldn't have done that.

"What'd you say?" Eddie said.

"I think he called you a punk," Ricky agitated.

"You call me a punk, man?" Eddie turned me around. I denied it.

"I heard him, dude. He say you are a punk-ass *puto,*" Ricky continued to exhort.

The fist came at me so fast, I don't even recall how Eddie looked when he threw it. I found myself on the ground. Others in the school had gathered around by then. When a few saw it was me, they knew it was going to be a slaughter.

I rose to my feet—my cheek had turned swollen and blue. I tried to hit Eddie, but he backed up real smooth and hit me again. Ricky egged him on; I could hear the excitement in his voice.

I lay on the ground, defeated. Teachers came and chased the boys out. But before Eddie and Ricky left they yelled back: "José ain't nothing, man. You ain't nothing."

Anger flowed through me, but also humiliation. It hurt so deep I didn't even feel the fracture in my jaw, the displacement which would later give me a disjointed, lopsided, and protruding chin. It became my mark.

Later when I told Rano what happened, he looked at me and shook his head.

"You didn't have to defend me to those dudes," he said. "They're assholes. They ain't worth it."

I looked at him and told him something I never, ever told him again.

"I did it because I love you."

I began high school a *loco,* with a heavy Pendleton shirt, sagging khaki pants, ironed to perfection, and shoes shined and heated like at boot camp.

Mark Keppel High School was a Depression-era structure with a brick and art-deco facade and small, army-type bungalows in back. Friction filled its hallways. The Anglo and Asian upper-class students from Monterey Park and Alhambra attended the school. They were tracked into the "A" classes; they were in the school clubs, they were the varsity team members and lettermen. They were the pep squads and cheerleaders.

But the school also took in the people from the Hills and surrounding community who somehow made it past junior high. They were mostly Mexican, in the "C" track (what were called the "stupid" classes), and who made up the rosters of the wood, print, and auto shops. Only a few of these students participated in school government, in sports, or in the various clubs.

The school had two principal languages. Two skin tones and two cultures. It revolved around class differences. The white and Asian kids (except for "barrio" whites and the handful of Hawaiians, Filipinos, and Samoans who ended up with the Mexicans) were from professional, two-car households with watered lawns and trimmed trees. The laboring class, the sons and daughters of service workers, janitors, and factory hands, lived in and around the Hills (or a section of Monterey Park called "Poor Side").

The school separated these two groups by levels of education: The professional-class kids were provided with college-preparatory classes; the blue-collar students were pushed into "industrial arts."

Creating a School to Meet the Needs of Poor Children

Loretta J. Brunious

This essay includes the comments of African American students about how they would create a school if given the opportunity. Their answers point to some of the problems in their existing school environments. From their comments, how would individual/cultural theorists and structural theorists explain their current school environments? It is possible to

SOURCE: Loretta J. Brunious, *How Black Disadvantaged Adolescents Socially Construct Reality* (New York: Garland Publishing, 1998), pp. 113–116. Reproduced by permission of Routledge, Inc., part of the Taylor & Francis Group.

learn effectively in substandard school environments? Why or why not? Compare this selection to the two prior selections. What common themes can you take from all three regarding schooling and poverty?

PERCEIVED FUTURE OF SCHOOLING IN RELATION TO THE COMMON SENSE WORLD

The students are given an opportunity to imagine and create a school that will meet their unique needs. Their answers represent a range and type of personally constructed realities for their age and stage of development in relation to their perceived social reality. Asked what type of school would they plan, they provide a rich range of responses.

QUESTION: If you plan a school the way you wanted it to be, what would it look like? I mean inside and out, what would it have? Would it be just like this school? Or would it be completely different? How would your school be?

SHAS: My school would be completely different. My school would be rebuilt. I would have marble and just nice desks. I would have everything, the walls would be redone. The windows would be tinted. The inside would be able to see outside but they couldn't see in. I would get some new wardrobes for the dance class, new cap and gowns, new choir gowns, new pianos, new desk, tables, chairs, books, cabinets. Everything. Get them a lunch lady who can cook, somebody to clean the school right. Mop the floor every night. I would find some teachers that would teach, make teaching fun, not sit there boring. Tired of them asking you to read this, read that, answer the questions.

L. LEE: It won't have all this graffiti and stuff on it, written on the wall and stuff. And to pass you have to be in your right level because they just pass them because they want to get rid of them. I know you can't stop the rats and roaches. It won't have no windows like this, broken and written all over.

JASON: It would be big. I have a big teachers parking lot and big playground for the kids and a playground for the pre-school and kindergarten. And for the upper grades they would have a tall building and for first to fifth grade there would be a small part of the building. First of all I would not have any mean teachers. I would pick-out every teacher. I would see what each teacher is doing. I try to find out what is their background. And then if they can do it perfect, I pick 'em. I have a swimming pool. A place where kids can go to learn more, like on the weekends. On the weekends the kids can come to school and do some more work, extra work. The principal would not be mean. They would be rules. They would have educational videos. And they would be draw art, culture club, dance, a lot of stuff. A lot of activity things. There would be south part of the building where the doors would be sound proof, so nobody can hears but the people that are in the dance room.

KEIS: First we'd have good food in the cafeteria. Plenty of books, desks. Make sure walls had no rats and roaches, no writing. Like if we had broken windows, we get them fixed right away. Heat would be working. Walls won't have holes in them and stuff. I'd let the kids talk in the lunchroom. But they couldn't get rowdy.

JOS: I get it rebuilt. Make it look nicer, decent. And make some new teachers come in here that feel good about the students.

G.T.: It would be like this school here, but, you know it would be cleaner and nobody in a gang. I would just throw them out. They can't come to this school. Who ever I catch fighting, no matter who right or wrong, they be thrown out too. We would have twenty-four hour security. That way nobody could write on the school.

NIC: It would be good. I would take out all the gangs. Good teachers, clean it up a little bit. It would be no gang bangers writing on the walls, there names or nothing.

SAM: I probably have it in the suburbs and stuff. Cause city kids more wild and more gangs out here. They have some in the suburbs but not many. Have better inspection and stuff. Try to keep it up, clean and all. Have nice teachers, nice attitudes.

MARY: It would be a better paying job, it wouldn't have no roaches in the lunchroom or classroom and it would be clean. It would be big enough and have all the equipment it need for science, social studies and all that. The teachers would be strict. Cause some of these kids they don't care. They just plain bad. They would be nice sometimes but they would be strict when it's learning time. The curriculum would be the same.

MAY: It would have a lot of activities. It would have blacks and whites and then it would give a good education. And I would support my school. I would try to get the teachers to teach more about our school and gangs and stuff. And don't start they when they get in eighth, like when they in first grade. Try to talk about them and try to talk to them about gangs and stuff and show them how people end up. And I use most of my brothers for examples.

THOM: I wouldn't want all black teachers. You be around a whole lot of black people, be a whole lot of arguments. A special place for disabled and handicapped kids, they have a special side of the school and then have K through eighth on the other side. Social studies would be different. Because it would be about black people.

Again, the students' remarks are insightful and rich. Their sense of need and want do not diverge: each of these children would escape the decay and degradation of their environment if they could. They would each design a place and a set of adults who could help them accomplish this extraordinary dream. The dream school stands in stark contrast to their environment: it is "rebuilt," "marble," "walls that are redone," and "tinted windows" so that they can "see out." This school has "no graffiti;" it has materials, equipment, "a lunch lady who can cook," and somebody who can "mop the floor every night." Here, the kids can come "on the weekends to do extra work," but "there would be rules." Mostly, "new teachers would feel good about the students." When these children speak of their reality—the omnipresent gangs, violence, and decay—they are tough in a world of constant conflict, a "war zone" populated by unfeeling and uncaring adults. But given the chance to dream, they know with certainty what kind of people and environment they need to realize themselves through education.

9

Work and Working

The jobs held by those in poverty or near poverty involve doing society's "dirty work." Some people in these jobs are migrant farmworkers who follow the harvests, picking fruit and vegetables; workers in slaughterhouses who process meat, poultry, and fish; dishwashers in restaurants; and maids who clean rooms in hotels. These and other low-end jobs involve low pay (some pay below the minimum wage), and few or no benefits, and some involve a high risk of injury from the activity itself or from exposure to toxic chemicals.

WAGES

About 20 percent of the nation's poor families have at least one family member holding a full-time job, yet they remain poor (Pollin and Luce, 1998). There are at least two related reasons for this anomaly. First, the federal minimum wage, despite occasional raises, has not kept up with inflation, showing a loss of more than 30 percent in purchasing power from 1968 to the present. Second, at the current minimum wage of $5.15, someone working full time for 50 weeks earns only $10,300 a year, which is far below the poverty threshold of $13,738 for a family of three or $17,603 for a family of four. It takes an hourly wage of more than $8.25 for a full-time worker to earn $17,603 a year, yet 21 percent of men and 34 percent of women earn less than this, as do 26 percent of whites, 34 percent of African Americans, and 45 percent of Latinos (Sharpe, 2001).

ALIENATION

Alienation refers to the separation of human beings from one another, from themselves, and from the products they create. Worker alienation occurs when workers are controlled in the work setting, when they receive no satisfaction and personal fulfillment from the work they do, when their work is not appreciated by others, when their work is repetitious, and when they are not paid a fair wage for their efforts. Clearly, those at the bottom of the job hierarchy are subject to these manifestations of alienation. Therefore, many feel a profound dissatisfaction and resentment against their employers and society. At a personal level, this may lead workers to join together in a union or other collective group to improve working conditions (**agency**), or it may manifest itself in absenteeism, work slowdowns, and alcohol or other drug abuse. Studs Terkel, in introducing his book *Working,* summarized the personal impact of alienating work:

> This book, being about work, is, by its very nature, about violence—to the spirit as well as the body. It is about ulcers as well as accidents, about shouting matches as well as fistfights, about nervous breakdowns, as well as kicking the dog around. It is, above all (or beneath all), about daily humiliations. To survive the day is triumph for the walking wounded among the great many of us. (Terkel, 1975:xiii).

RISK

Those who work for low wages often do dangerous work. Migrant farmworkers, for example, are exposed to herbicides and pesticides that are sprayed on the fields where they work. The water they drink may be contaminated, and their living conditions are often unsanitary. As a result these workers have a life expectancy 30 years below the national average. Workers in packing plants and chicken and fish processing plants use sharp knives for gutting carcasses. They often receive cuts and are susceptible to repetitive motion syndrome. In other low-wage jobs, workers may be exposed to nuclear radiation, chemical compounds, coal tars, dust, and asbestos fibers.

SWEATSHOPS

A **sweatshop** is a substandard work environment where workers are paid less than the minimum wage, where workers are not paid overtime premiums, and where other laws are violated. Although sweatshops occur in various types of manufacturing, they occur most frequently in the garment industry. The Labor Department estimates that more than half of the 22,000 U.S. sewing businesses are sweatshops (Branigan, 1997). Most workers in these establishments are Latina and Asian immigrant women who are paid much below the minimum wage, receive no benefits, and work in crowded, unsafe, and stifling conditions.

The essays in this chapter show in vivid detail what it is like to do society's dirty work. Compare their working conditions and pay with middle-class jobs and executive jobs. What is your reaction to the fact that the average chief executive officer of a major corporation makes 476 times more than the average worker in that corporation? Consider, too, the differences in benefits, including "golden parachutes" worth many millions if an executive is fired and a pink slip if an assembly line worker is fired. What is fair when it comes to wages and benefits? What should be the role of unions in determining what is fair compensation? Should all workers receive a "living wage"?

NOTES AND SUGGESTIONS FOR FURTHER READING

Bernhardt, Annette (2000). "The Wal-Mart Trap." *Dollars and Sense,* Number 231 (September/October):23–25.

Branigan, William (1997). "Sweatshops Are Back." *The Washington Post National Weekly Edition* (February 24):6–7.

Chang, Grace (2000). *Disposable Domestics: Immigrant Women Workers in the Global Economy*. Cambridge, MA: South End Press.

Chatterley, Cedric N., and Alicia J. Rouverol with Sephen A. Cole (2000). *"I Was Content and Not Content": Story of Linda Lord and the Closing of Penobscot Poultry.* Carbondale: Southern Illinois University Press.

Duncan, Cynthia M. (1999). *World's Apart: Why Poverty Persists in Rural America.* New Haven, CT: Yale University Press.

Ehrenreich, Barbara (2000). "Maid to Order: The Politics of Other Women's Work." *Harper's* 300 (April):59–70.

Ehrenreich, Barbara (2001). *Nickel and Dimed: On (Not) Getting By in America.* New York: Metropolitan Books.

Heinz, James, Nancy Folbre, and the Center for Popular Economics (2000). *The Ultimate Field Guide to the U.S. Economy*. New York: The New Press.

Levy, Frank (1998). *The New Dollars and Dreams: American Incomes and Economic Change*. New York: The Russell Sage Foundation.

Mishel, Lawrence, Jared Bernstein, and John Schmitt (2000). *The State of Working America 2000–2001*. Ithaca, NY: Cornell University Press.

Newman, Katherine S. (1999). *No Shame in My Game: The Working Poor in the Inner City*. New York: Alfred A. Knopf and the Russell Sage Foundation.

Pollin, Robert, and Stephanie Luce (1998). *The Living Wage: Building a Fair Economy*. New York: The New Press.

Rubin, Lillian B. (1994). *Families on the Faultline: America's Working Class Speaks About the Family, the Economy, Race, and Ethnicity*. New York: HarperCollins.

Sharpe, Rochelle (2001). "What Exactly is a 'Living Wage'?" *Business Week* (May 28):78–79.

Sonneman, Toby R., and Rick Steigmeyer (1992). *Fruit Fields in My Blood: Okie Migrants in the West.* Moscow, ID: University of Idaho Press.

Terkel, Studs (1975). *Working: People Talk about What They Do All Day and How They Feel about What They Do.* New York: Avon Books.

Looking for Work, Waiting for Work

Toby F. Sonneman

This selection is an in-depth look at the life of migrant fruit pickers. These workers provide a valuable service to our economy, and yet they receive little in return. Explain their working conditions from individual/cultural and structural perspectives. What are the disadvantages and advantages of migrant work? What steps could migrant workers take to improve their lives and working conditions?

In late April, the early variety of cherries in the orchards around Stockton, California, begins to turn from pale green to reddish-pink, and the roads are full of pickers looking for work. Often migrant fruit pickers return, like migratory birds, to the orchards where they have worked every spring for years. For those pickers who find themselves without a job to return to, there are days ahead of driving back roads looking for work, talking to growers, contractors, and other pickers for leads or connections to a job.

Even for those lucky enough to have an annual job to come to, there may be a long period of waiting before the work actually begins. In a journal I kept years ago, I described the situation of waiting for work in a migrant camp:

27 May

Arriving in Lodi yesterday to a full camp. Everything seems much the same as last year—the old crew with a few new additions. Some of the people have new rigs—usually larger or newer trailers and trucks. The kids have grown bigger. The ground is still that dusty burnt ash. The smoky dust blows into the trailers, turns the children coal gray, and covers everything with a film of dirt.

It's hot here, in the nineties, but the evening breeze and the shade of a few tall trees make it more tolerable.

We expected the mood of the camp to be dismal. Almost everyone has been camping here for a month or a month and a half, yet there's been only half a day of work. The early cherries were rained out. Now we're waiting for the late cherries to start—a rather hopeless prospect also. Most of the blossoms came off in the rain, leaving hardly any cherries on the trees. As if that weren't bad enough, we have a bigger-than-average crew. Soft-hearted Otis has let on a lot of extra people—friends and relatives of the crew. He planned on needing them in the early cherries but since that fell through they've all waited for the late cherries.

A combination of desperation and excitement pervades the camp full of pickers waiting for the job to begin. Each day of waiting is filled with anxiety and anticipation but little activity. There's not much to do in the surrounding community that doesn't involve spending money, and money is in short supply for migrants this time of year. They've spent what they have in travel, and they may have gone for months without work. Consequently this time of waiting becomes a time to socialize with other pickers in the camp, and the focus of the conversation is often speculation about the crop.

Among Okie migrants, it's the men who seem to have more time to socialize, while the women—in traditional roles—cook, clean, shop, and watch the children. Men especially tend to define the limbo time of waiting with a particular ritual each morning.

After early breakfasts, men drift out of their trailers, coffee cups or cigarettes in hand. The trailers are parked close to-

SOURCE: Toby F. Sonneman, *Fruit Fields in My Blood: Okie Migrants in the West* (Moscow, ID: University of Idaho Press, 1992), pp. 53–61. Used by permission of the publisher.

gether on a cleared patch of land surrounded by orchard. The dirt road through the orchard borders the migrant camp on one side. The men walk casually across the road to the orchard, greeting each other with a nod of the head. "Mornin', Otis. Ya think we're ever goin' to get ta pick this fruit?"

The men form a small circle on the edge of the orchard. There are six or seven of them, their ages ranging from twenty-five to sixty. Rick and I hang back along the edge of the circle, listening and occasionally participating. Although I am conspicuously the only woman in the group, the men take little notice of me. They examine the unripe fruit carefully, placing clusters of the small pale cherries in their hands, discussing their condition minutely, as if they could will them to ripen sooner. "They still look pretty green to me. I reckon it's ten days away," Otis comments. "Well, now I wisht I'd stayed in Arizona. There ain't much goin' on there, but at least I could make enough to buy groceries."

One of the younger men leans against the trunk of a tree. He has the look of another time: his face is pale and narrow, with a long forehead and deep-set hazel eyes, topped by a shock of blond curls. Striped overalls and a worn work shirt cover his thin body. "Well, you know a fruit tramp," he grins, hooking his thumbs in his overall pockets. "When he decides to go, he's gone! We'd be pickin' oranges down there in Porterville, and someone'd say, 'Cherries are a good four weeks off, no use in goin' up north.' Next mornin' they'd change their mind and they'd be gone. Sometimes they'd go up there for a day and turn right around and come back, and then go up there again. Yeah, about six weeks before the cherries start, there's no tellin' what will happen!"

"It's funny," agrees another, addressing his explanation to Rick. "We only pick cherries about forty days a year, but we call ourselves cherry pickers."

"The cherries is all there is now." A stocky man squats near the ground, drawing lines in the dust with a twig. "If you don't' make it in the cherries, you ain't gonna make it. Though there's some that make it in the apples too. But you can't hit it in the cherries and then try to make the same kind of money in the prunes and peaches and apples and oranges all through the year. If you did, you'd be dead by the time you were twenty-five!"

"Well, we make pretty good in the cherries," another man says. "But by the time you figure there's eight people working, and the whole rest of the year when we can't make that much, I doubt if I make more'n twenty dollars a day."

"I ain't too sure we're going' to make it at all this year," the young man in overalls says wryly. "We might just be standin' around the welfare office if they don't ripen up here real quick. You know, they say there's only a dime's difference between the man that works and the man that don't work—and the one that don't work is the one with the dime!"

As everyone chuckles at that last line, a truck pulls up beside the group and the grower, a young man with a tanned face, leans his head out of the window. One of the older pickers takes a few steps out of the group to talk with him. "They're still awful green. Just not ripening up in this cool weather," the picker says, shaking his head.

"Yeah, I'm hoping we don't get any more rain now," comments the grower. "I reckon they're ten days off. Well, the field man will be out Wednesday to let us know." The grower waves goodbye and the truck ambles down the orchard road.

The men stand around a little awkwardly, their ostensible purpose of evaluating the cherries now gone. Then they start to disperse, making excuses as they move away. "Well, I guess I should see if Marge is ready to go into town," says one. "Yeah, I got to look at that truck of mine. That starter's been actin' up," says another. They seem disheartened as they walk back across the road. They know there won't be any work for at least ten days, and there's nothing to do but wait. The cherries won't be ripe up north for weeks to come. But often there are other jobs in the area if one has the patience and persistence to pursue them.

Going to search for a job can be a time-consuming and discouraging task. "Come back around five A.M. on Thursday and I'll see if I can put you on," a grower is apt to say. "If some of my pickers don't show up, I'll have a ladder for you." Pickers who really want the job must rise early, get ready for the day's work and drive—often ten miles or more. More often than not they will be told, "Sorry, I'm full up."

Before most pickers will even ask about a job they will spend many hours and gallons of gas driving around checking out the orchards. They're looking for large cherries, trees loaded with fruit, and trees spaced apart so their limbs aren't tangled together. Preferably, the trees are well pruned and a picker can reach all the cherries with a ten- or twelve-foot ladder.

Often a group of relatives or friends camped out together at a pickers' camp will go look for another job together. Since they may come to an orchard camp weeks before the work there is scheduled to begin, they need other work to fill in the time.

One time when we were staying in a migrant camp with about a week before our next job was to start we set out with several other pickers to look for another job. Like them, we couldn't miss days of possible work during the short-lived cherry season. Ruby, an older woman, had said she could use her "connection" to a grower to get us all jobs. But before we committed ourselves to asking for a job, we wanted to check out the orchard surreptitiously to be sure that we could all do well there. Crammed into a car with six other people, we drove several miles down back roads lined with orchards, until Ruby directed us to turn down a dirt driveway. We stopped in the middle of the orchard. The orchard was quiet—the cherries weren't quite ripe enough for picking so no one else was there. We swung open the four car doors and got out to congregate for a few minutes under the closest tree.

"Man, these trees are really loaded!" exclaimed Walter. His brother Daniel chewed thoughtfully on a cherry. "They sure are roped on there alright." The men drifted off to another tree, picking up branches to feel how heavily they were set with cherries and eating an occasional cherry to assess its size and ripeness.

Meanwhile the women remained near the car. Darlene scrutinized the first tree with the observant eye of experience. "They're loaded alright, but look at the size of 'em. I bet a lot of these cherries won't even make twelve-row!" she said critically.

Ruby agreed. "There ain't nothin' worse than pickin' those little cherries. You pick and pick and your bucket never does seem to get full!"

The men returned, meandering through the orchard investigating a clump of cherries carefully. "There's some pretty loaded trees out there," Daniel concluded. "It looks like a good job to me."

The women disagreed. "You know how those little cherries just pack down in your bucket," Darlene reminded us.

Walter interrupted, saying "Let's just go on up there after awhile and ask about it. We can't be no worse off than we are now, with no job." Despite the pickers' attempt to be selective and discerning when looking for a job, the bottom line was often what was available when they badly needed a job. Because the best money is in the cherries and then in the apples, there is intense pressure to work every available day during these harvests. There is plenty of dead time in between harvests and in the winters—pickers want to insure that they have a minimum of days off during the harvests.

Besides driving around to look for work, pickers usually call ahead to a farmer they have worked for to confirm their next job. They also rely on the grapevine—word-of-mouth reports of crop conditions and jobs, which are often completely without truth. One year a rumor circulated that our next job up north—a job that many people were counting on to make the bulk of their money in the cherries—would not be hiring any of us. The rumor was that the farmer had fired the contractor and had hired a whole crew of "wetbacks." Everyone was upset about the possibility and the rumor persisted for several days until phone calls were made to the grower, who denied the rumor.

The spread of false rumors about crop conditions is also commonplace. "I hear that hailstorm in the Tri-Cities wiped out the whole crop." "They say the trees are so loaded up in Brewster that one person can make eighty dollars by noon." These are examples of both negative and positive rumors we've heard about crops. Often such reports have a grain of truth in them but are very exaggerated. If a picker tells you about a bumper crop, it's likely that it's a good crop although it may not be as good as you were led to believe. Similarly, impressive stories about how much money a picker made a day picking cherries are usually misleading. Although such stories can be true, they usually represent one long day of incredibly hard effort by an experienced picker in the best fruit. Pickers realize the distortions such stories can lead to, but they take pride in their picking skills and want to tell about their abilities. Unfortunately, farmers often use these unrepresentative high figures to try to prove that pickers are making plenty of money—in their view, too much.

Exaggerations or distortions also occur about the length of a job and when it will begin. If a picker says it's a ten-day job, it usually lasts a week; a week-long job is usually finished in five days. If picking is supposed to start Tuesday, more often than not it will be Wednesday or Thursday before it actually begins.

Pickers can also be misled by farmers and contractors, who rely on field men for the go-ahead to begin picking. They may be told that a job will begin on a certain date and even leave one job early to travel to the next one, only to have to wait for days before the work begins.

Even under the best circumstances, when a picker has secured and begun a good job, there are irritating delays, layoffs because of rain or unripeness of a variety of fruit (for example, in the same cherry orchard, Bings, Lamberts, Vans, and other varieties will ripen at staggered times). A typical work record I kept during a ten-day period at a cherry orchard in Kennewick, Washington, illustrates this frustrating fact of agricultural work:

June 8—Worked 5 A.M. to 2:30 P.M., averaged 4 boxes a tree.
June 9—Got laid off at 1:00.
June 10—Rained out at 10:00.
June 11—Laid off.
June 12—Laid off.
June 13—Got a good set after half a day, 5 A.M.–2:30 P.M.
June 14—Doing pretty good but he called us out at 1:00 P.M.!
June 15—Finally made our $100! 5 A.M.–3 P.M.
June 16—Again! Started in Vans about noon.
June 17—Took until 1:30 to finish. Windy and dusty!

The record shows that within the ten days we spent on the job, we had only four full days of work. Of course, even on the three days that we had to quit at one o'clock, we had already put in eight hours of work, but we didn't consider that a full day of work in the cherries. Because of the long days during the cherry harvest and the pressure to make most of our money then, we needed to work nine or ten hours to "make our quota." That quota, which may differ from picker to picker and job to job, is a personal expectation of how many boxes will be picked each day. Countless factors can erode such expectations. Because rain causes picked cherries to rot in the box, pickers are called out of the orchard in anything more than a light drizzle. Moving to other sections of the orchard, finishing one variety of cherries and starting another takes large chunks of the pickers' time, for which they are not compensated. Finally, bad trees—ones with too few cherries or cherries that don't come off easily; or ones with unpruned or tangled limbs—can cause pickers to not meet their quotas.

But in spite of all the frustrations, delays, and disappointments, there is nothing so satisfying to a picker as a good day of fruit picking. It's a rare day when the weather is perfect, not too cool or too hot, and a promising row of loaded trees stretches out ahead. The camaraderie of the crew is pleasant and the boxes of cherries seem to fill almost effortlessly with plump fruit. These days are rare even in the best harvest, but they are always what a picker hopes for at the beginning of the fruit run and remembers long after the last fruit of the season is picked.

Chicken Run

An Interview with Linda Lord

This interview takes place in Belfast, Maine, with a worker in a poultry-processing plant. As you read this selection, think about the phrase "society's dirty work." What kind of job conditions did Linda Lord face? After the closing of the factory, what does Linda Lord take away from her experience? In other words, what do you think are her life chances? What role can unions play in factory work such as this? Take this selection and first analyze it from a non-sociological perspective; then analyze it again using your sociological imagination.

Linda Lord began gutting chicken at Penobscot Poultry, a small processing plant in Belfast, Maine, shortly after she finished high school in the late 1960s. "The machine got, oh, maybe 65 to 70 percent of the birds," Lord says, "and the rest I had to do." Her saga, told by herself in the course of six interviews, is at the heart of "I Was Content . . . And Not Content," a story of low-tech work in the new high-tech economy. Belfast, the authors say, did little to attract new industry or retrain workers, instead setting up tax shelters for financial service companies like MBNA America. "Under what conditions does our economic system undermine our efforts to create healthy and stable communities?" they ask.

(The following abbreviations have been used: AR for Alicia J. Rouverol; CC for Cedric N. Chatterley; SC for Stephen A. Cole; and LL for Linda J. Lord.)

SOURCE: Cedric N. Chatterley and Alicia J. Rouverol, with Stephen A. Cole. *"I Was Content and Not Content": The Story of Linda Lord and the Closing of Penobscot Poultry* (Carbondale, IL: Southern Illinois University Press, 2000), pp. 1–25. Used by permission of the publisher.

March 1, 1988; Gallagher's Galley, Brooks, Maine

SC: Linda, were you born in Brooks?

LL: No, I was born up in Waterville at the old sisters' hospital. But I've always lived here in Brooks. . . . I went to my freshman year at Moss Memorial High School, and then they were building the area school up on Knox Ridge, and I started my sophomore year up there and graduated from Mount View [in 1967]. After high school I went right into the hospital for an operation, and I wasn't supposed to work for a year; and come September, I got edgy and I started working a short time over at the [Unity] hatchery before I went down to the plant.

SC: What did you do?

LL: I de-beaked chickens, sexed them, injected them, de-toed them, de-beaked them, you name it I did it. [laughs]

SC: What is de-beaking?

LL: De-beaking is burning part of the bill off so they don't peck each other as they get bigger. . . .

SC: And then you learned to sex chickens, too. We hear that's rather a specific and actually sort of hard-to-learn task. Did you like working out there?

LL: It was all right. It was a job. At that time I was young, and I wanted to stay close to home on account of my mom being ill. And I wasn't there very long before they transferred me down at the plant because things were getting slow. It was either they were going to be laying off, or I had a chance to go up to the plant. I had looked around for other jobs, and I figured, well, I had been working with the company, I'd stay with them. So I went down to Penobscot.

SC: Let me back up a little bit. Did you go to the hatchery to work initially because your dad had also worked in the poultry industry?

LL: My father had been a pullorum [poultry blood] tester for a good number of years with the University of Maine. He did work for Penobscot and at the time for Maplewood [closed 1980]. Of course, during the summers, I worked for him testing birds. So I got to know quite a lot. Of course, we raised birds, too; we had laying hens for Maplewood so I'd grown right up. But no, that wasn't why I went over to the hatchery. It was, at that time, that was just about the only place that was hiring, you know. And I wanted a job. I was getting edgy not doing anything, even though I was supposed to stay out a year and not even work.

SC: Did other people you know from Mount View also hook up with Penobscot for a job?

LL: Not too many, no. As a matter of fact, a lot of the guys in my class were wiped out in Vietnam. I think we lost four or five. And the ones that did come back were disabled or not with it or either freaked right out on dope, you know.

SC: That's a large number of people from one high school. Well, probably your graduating class wasn't all that large in the end.

LL: I think it was a little over—maybe like a hundred and twenty in our class. I can't remember now, but I've got it on the back of my diploma. . . .

SC: So they transferred you down to Penobscot. What did you start doing there?

LL: Transferring. That's hanging [poultry] from the "New York" room after the feathers have been taken off and they've gone through the foot cutter—they drop down to the belt. I was working on Line Two, which did big birds and small birds. . . . But I kept breaking the skin away from my nails and getting blood poisoning streamers going right up my arm, so that's when I signed up for the sticking job. And the pay was a lot better. At that time I was going through a divorce, so I was out to get as much money as I could to support myself.

SC: What specifically was your job then?

LL: Just putting the bird into the shackle going out on the eviscerating line, after the feet have been cut off. They drop down on a belt which came to you, and you just would put them in a shackle so they could go out to be opened up and their intestines drawn out, and get your heart and liver and so forth out of the bird.

SC: So it's rough on your hands. How do you think that caused blood poisoning?

LL: Well, you take that infection that would set in, even though you'd try washing your hands good. And eventually it would get into your blood system and pretty soon, if you didn't take antibiotics or something, you'd have red streamers going right up your arm. And you'd take what helped to cut the feathers so they would pick better—the solution they added with the water [to remove the feathers]—and you'd take it with the grease and stuff, and that would cause you more trouble and make your skin peel right off your hands, too. My hands now are even sensitive to hot and cold where I was transferring, and it's been a number of years.

SC: So how long did you do that?

LL: About four or five years, then I signed right up for the rest of the time to go right on sticking.

SC: Were there a fair number of other people who work on the lines who have had these kinds of problems?

LL: Everybody's had problems that's been there. A lot of people have had warts come out on their hands because of handling chicken. A lot of people have had tendonitis, you know. That work is hard work down there. And they've had blood poisoning, you name it. Or you get, well—what do they call it? You have like a rash break out all over you from the chickens, too, which eventually will blister right up with little pus sacks, and the skin will peel right off your hand, and they call it—"chicken poisoning" is what they call it. And I've seen people get it from wearing rubber gloves.

SC: Wow. Were you ever able to get any satisfaction from the company in terms of better conditions so that wouldn't happen? You had health benefits, right?

LL: Yeah, and if you had blood poisoning or anything, they'd send you right down to the hospital to try to get it cleared up. At that time they would put you on easy jobs until you either got over tendonitis or the rash on your hands and so forth. But as workmen's comp passed a few laws and stuff, then they had people just stay right out of work. They didn't want you even in there doing light work.

SC: Did that mean that if people had a problem, they were less likely to tell the company if that meant they now had to go home?

LL: A lot of people would stick it out and not say much. Because you had to be out over three days in order to draw on workmen's comp. If you were out maybe one or two days, you just couldn't get anything—you lost a couple days work. So a lot of people just kept going.

SC: So that job you had four or five years, and that's a total of how many years that you were at Penobscot?

LL: Twenty years in all with the company.

SC: Twenty years. So after that [transferring] it was into the sticking room or the "blood tunnel"?

LL: The sticking room, the "blood tunnel," or what I called the "hell hole," where they had so much blood. [laughter] No one wanted to come in there when you were in there. You were just by yourself until you got done work.

When I first started out sticking, we didn't have any machinery in there then, except for just the stunners. And that first stunner that made the bird's head hang down is where we usually sat—two of us. There were two stickers in there and we had to do every other bird running right in a full line. And then about

'79 . . . they were thinking about increasing the production down there. So they went into the sticking machines, which at first didn't pan out very good. But after a while, about a half a year, they got it straightened out so it would do a pretty good job. And then you had to back up the machine.

SC: So initially the stunner stunned the bird, but you had to stick them.

LL: Right. Grab it, take a knife and cut the vein right in by the jaw bone.

SC: So, by stunning, essentially that means that the bird was in shock? Were the birds all pretty much dead by the time they got to you?

LL: No, they were still, you know, flopping their last flop before they died. That's why I was more or less, as you could see—with the rain kerchief, I was more or less covered right up so I wouldn't get too bloody. . . . The machine got, oh, maybe sixty-five to seventy percent of the birds, and the rest I had to do. . . .

SC: Did you make more money working in that job?

LL: Yeah, because that was top pay. I mean I got the same as the trailer truck drivers did. The last of it then was $5.69 an hour, which was more than what the people were getting on the line. Maybe five, ten, fifteen, or twenty cents more—because each job through that plant you had different wages. . . . The trailer truck drivers, or straight job drivers, and stickers and weighers got the same wages—$5.69 an hour.

At a Slaughterhouse

Charlie LeDuff

This selection is by a New York Times *reporter who signed up for work in the slaughterhouse. While not poor himself, he took a job at the slaughterhouse so he could describe the working conditions and how race structures the work environment there. What did he find in terms of the intersection of work and race? What is the situation of the illegal immigrants there? How do divisions by race actually benefit the employers of a slaughterhouse? What might a union do for these workers? Why do you think they voted down having a union?*

Tar Heel, N.C.—It must have been 1 o'clock. That's when the white man usually comes out of his glass office and stands on the scaffolding above the factory floor. He stood with his palms on the rails, his elbows out. He looked like a tower guard up there or a border patrol agent. He stood with his head cocked.

One o'clock means it is getting near the end of the workday. Quota has to be met and the workload doubles. The conveyor belt always overflows with meat around 1 o'clock. So the workers double their pace, hacking pork from shoulder bones with a driven single-mindedness. They stare blankly, like mules in wooden blinders, as the butchered slabs pass by.

It is called the picnic line: 18 workers lined up on both sides of a belt, carving meat from bone. Up to 16 million shoulders a year come down that line here at the Smithfield Packing Co., the largest pork production plant in the world. That works out to about 32,000 a shift, 63 a minute, one every 17 seconds for each worker for eight and a half

SOURCE: Charlie LeDuff, "At a Slaughterhouse, Some Things Never Die," *New York Times* (June 16, 2000). Used by permission of the publisher.

hours a day. The first time you stare down at that belt you know your body is going to give in way before the machine ever will.

On this day the boss saw something he didn't like. He climbed down and approached the picnic line from behind. He leaned into the ear of a broad-shouldered black man. He had been riding him all day, and the day before. The boss bawled him out good this time, but no one heard what was said. The roar of the machinery was too ferocious for that. Still, everyone knew what was expected. They worked harder.

The white man stood and watched for the next two hours as the blacks worked in their groups and the Mexicans in theirs. He stood there with his head cocked.

At shift change the black man walked away, hosed himself down and turned in his knives. Then he let go. He threatened to murder the boss. He promised to quit. He said he was losing his mind, which made for good comedy since he was standing near a conveyor chain of severed hogs' heads, their mouths yoked open.

"Who that cracker think he is?" the black man wanted to know. There were enough hogs, he said, "not to worry about no fleck of meat being left on the bone. Keep treating me like a Mexican and I'll beat him."

The boss walked by just then and the black man lowered his head.

WHO GETS THE DIRTY JOBS

The first thing you learn in the hog plant is the value of a sharp knife. The second thing you learn is that you don't want to work with a knife. Finally you learn that not everyone has to work with a knife. Whites, blacks, American Indians and Mexicans, they all have their separate stations.

The few whites on the payroll tend to be mechanics or supervisors. As for the Indians, a handful are supervisors; others tend to get clean menial jobs like warehouse work. With few exceptions, that leaves the blacks and Mexicans with the dirty jobs at the factory, one of the only places within a 50-mile radius in this muddy corner of North Carolina where a person might make more than $8 an hour.

While Smithfield's profits nearly doubled in the past year, wages have remained flat. So a lot of Americans here have quit and a lot of Mexicans have been hired to take their places. But more than management, the workers see one another as the problem, and they see the competition in skin tones.

The locker rooms are self-segregated and so is the cafeteria. The enmity spills out into the towns. The races generally keep to themselves. Along Interstate 95 there are four tumbledown bars, one for each color: white, black, red and brown.

Language is also a divider. There are English and Spanish lines at the Social Security office and in the waiting rooms of the county health clinics. This means different groups don't really understand one another and tend to be suspicious of what they do know.

You begin to understand these things the minute you apply for the job.

BLOOD AND BURNOUT

"Treat the meat like you going to eat it yourself," the hiring manager told the 30 applicants, most of them down on their luck and hungry for work. The Smithfield plant will take just about any man or woman with a pulse and a sparkling urine sample, with few questions asked. This reporter was hired using his own name and acknowledged that he was currently employed, but was not asked where and did not say.

Slaughtering swine is repetitive, brutish work, so grueling that three weeks on the factory floor leave no doubt in your mind about why the turnover is 100 percent. Five thousand quit and five thousand are hired every year. You hear people say, They don't kill pigs in the plant, they kill people. So desperate is the company for workers, its recruiters comb the streets of New York's immigrant communities, personnel staff members say, and word of mouth has reached Mexico and beyond.

The company even procures criminals. Several at the morning orientation were inmates on work release in green uniforms, bused in from the county prison.

The new workers were given a safety speech and tax papers, shown a promotional video and informed that there was enough methane, ammonia and chlorine at the plant to kill every living thing here in Bladen County. Of the 30 new employees, the black women were assigned to the chitterlings room, where they would scrape feces and worms from intestines. The black men were sent to the butchering floor. Two free white men and the Indian were given jobs making boxes. This reporter declined a box job and ended up with most of the Mexicans, doing knife work, cutting sides of pork into smaller and smaller products.

Standing in the hiring hall that morning, two women chatted in Spanish about their pregnancies. A young black man had heard enough. His small town the next county over was crowded with Mexicans. They just started showing up three years ago—drawn to rural Robeson County by the plant—and never left. They stood in groups on the street corners, and the young black man never knew what they were saying. They took the jobs and did them for less. Some had houses in Mexico, while he lived in a trailer with his mother.

Now here he was, trying for the only job around, and he had to listen to Spanish, had to compete with peasants. The world was going to hell.

"This is America and I want to start hearing some English, now!" he screamed.

One of the women told him where to stick his head and listen for the echo. "Then you'll hear some English," she said.

An old white man with a face as pinched and lined as a pot roast complained, "The tacos are worse than the niggers," and the Indian leaned against the wall and laughed. In the doorway, the prisoners shifted from foot to foot, watching the spectacle unfold from behind a cloud of cigarette smoke.

The hiring manager came out of his office and broke it up just before things degenerated into a brawl. Then he handed out the employment stubs. "I don't want no problems," he warned. He told them to report to the plant on Monday morning to collect their carving knives.

$7.70 AN HOUR, PAIN ALL DAY

Monday. The mist rose from the swamps and by 4:45 A.M. thousands of headlamps snaked along the old country roads. Cars carried people from the backwoods, from the single and double-wide trailers, from the cinder-block houses and wooden shacks: whites from Lumberton and Elizabethtown; blacks from Fairmont and Fayetteville; Indians from Pembroke; the Mexicans from Red Springs and St. Pauls.

They converge at the Smithfield plant, a 973,000-square-foot leviathan of pipe and steel near the Cape Fear River. The factory towers over the tobacco and cotton fields, surrounded by pine trees and a few of the old whitewashed plantation houses. Built seven years ago, it is by far the biggest employer in this region, 75 miles west of the Atlantic and 90 miles south of the booming Research Triangle around Chapel Hill.

The workers filed in, their faces stiffened by sleep and the cold, like saucers of milk gone hard. They punched the clock at 5 A.M., waiting for the knives to be handed out, the chlorine freshly applied by the cleaning crew burning their eyes and throats. Nobody spoke.

The hallway was a river of brown-skinned Mexicans. The six prisoners who were starting that day looked confused.

"What the hell's going on?" the only white inmate, Billy Harwood, asked an older black worker named Wade Baker.

"Oh," Mr. Baker said, seeing that the prisoner was talking about the Mexicans. "I see you been away for a while."

Billy Harwood had been away—nearly seven years, for writing phony payroll checks from the family pizza business to buy crack. He was Rip Van Winkle standing there. Everywhere he looked there were Mexicans. What he didn't know was that one out of three newborns at the nearby Robeson County health clinic was a Latino; that the county's Roman Catholic church had a special Sunday Mass for Mexicans said by a Honduran priest; that the schools needed Spanish speakers to teach English.

With less than a month to go on his sentence, Mr. Harwood took the pork job to save a few dollars. The word in jail was that the job was a cakewalk for a white man.

But this wasn't looking like any cakewalk. He wasn't going to get a boxing job like a lot of other whites. Apparently inmates were on the bottom rung, just like Mexicans.

Billy Harwood and the other prisoners were put on the picnic line. Knife work pays $7.70 an hour to start. It is money unimaginable in Mexico, where the average wage is $4 a day. But the American money comes at a price. The work burns your muscles and dulls your mind. Staring down into the meat for hours strains your neck. After thousands of cuts a day your fingers no longer open freely. Standing in the damp 42-degree air causes your knees to lock, your nose to run, your teeth to throb.

The whistle blows at 3, you get home by 4, pour peroxide on your nicks by 5. You take pills for your pains and stand in a hot shower trying to wash it all away. You hurt. And by 8 o'clock you're in bed, exhausted, thinking of work.

The convict said he felt cheated. He wasn't supposed to be doing Mexican work. After his second day he was already talking of quitting. "Man, this can't be for real," he said, rubbing his wrists as if they'd been in handcuffs. "This job's for an ass. They treat you like an animal."

He just might have quit after the third day had it not been for Mercedes Fernández, a Mexican. He took a place next to her by the conveyor belt. She smiled at him, showed him how to make incisions. That was the extent of his on-the-job training. He was peep-eyed, missing a tooth and squat from the starchy prison food, but he acted as if this tiny woman had taken a fancy to him. In truth, she was more fascinated than infatuated, she later confided. In her year at the plant, he was the first white person she had ever worked with.

The other workers noticed her helping the white man, so unusual was it for a Mexican and a white to work shoulder to shoulder, to try to talk or even to make eye contact.

As for blacks, she avoided them. She was scared of them. "Blacks don't want to work," Mrs. Fernández said when the new batch of prisoners came to work on the line. "They're lazy."

Everything about the factory cuts people off from one another. If it's not the language barrier, it's the noise—the hammering of compressors, the screeching of pulleys, the grinding of the lines. You can hardly make your voice heard. To get another's attention on the cut line, you bang the butt of your knife on the steel railings, or you lob a chunk of meat. Mrs. Fernández would sometimes throw a piece of shoulder at a friend across the conveyor and wave good morning.

THE KILL FLOOR

The kill floor sets the pace of the work, and for those jobs they pick strong men and pay a top wage, as high as $12 an hour. If the men fail to make quota, plenty of others are willing to try. It is mostly the blacks who work the kill floor, the stone-hearted jobs that pay more and appear out of bounds for all but a few Mexicans.

Plant workers gave various reasons for this: The Mexicans are too small; they don't like blood; they don't like heavy lifting; or just plain "We built this country and we ain't going to hand them everything," as one black man put it.

Kill-floor work is hot, quick and bloody. The hog is herded in from the stockyard, then stunned with an electric gun. It is lifted onto a conveyor belt, dazed but not dead, and passed to a waiting group of men wearing blood-stained smocks and blank faces. They slit the neck, shackle the hind legs and watch a machine lift the carcass into the air, letting its life flow out in a purple gush, into a steaming collection trough.

The carcass is run through a scalding bath, trolleyed over the factory floor and then dumped onto a table with all the force of a quarter-ton water balloon. In the misty-red room, men slit along its hind tendons and skewer the beast with hooks. It is again lifted and shot across the room on a pulley and bar, where it hangs with hundreds of others as if in some kind of horrific dry-cleaning shop. It is then pulled through a wall of flames and met on the other side by more black men who, stripped to the waist beneath their smocks, scrape away any straggling bristles.

The place reeks of sweat and scared animal, steam and blood. Nothing is wasted from these beasts, not the plasma, not the glands, not the bones. Everything is used, and the kill men, repeating slaughterhouse lore, say that even the squeal is sold.

The carcasses sit in the freezer overnight and are then rolled out to the cut floor. The cut floor is opposite to the kill floor in nearly every way. The workers are mostly brown—Mexicans—not black; the lighting yellow, not red. The vapor comes from cold breath, not hot water. It is here that the hog is quartered. The pieces are parceled out and sent along the disassembly lines to be cut into ribs, hams, bellies, loins and chops.

People on the cut lines work with a mindless fury. There is tremendous pressure to keep the conveyor belts moving, to pack orders, to put bacon and ham and sausage on the public's breakfast table. There is no clock, no window, no fragment of the world outside. Everything is pork. If the line fails to keep pace, the kill men must slow down, backing up the slaughter. The boxing line will have little to do, costing the company payroll hours. The blacks who kill will become angry with the Mexicans who cut, who in turn will become angry with the white superintendents who push them.

10,000 UNWELCOME MEXICANS

The Mexicans never push back. They cannot. Some have legitimate work papers, but more, like Mercedes Fernández, do not.

Even worse, Mrs. Fernández was several thousand dollars in debt to the smugglers who had sneaked her and her family into the United States and owed a thousand more for the authentic-looking birth certificate and Social Security card that are needed to get hired. She and her husband, Armando, expected to be in debt for years. They had mouths to feed back home.

The Mexicans are so frightened about being singled out that they do not even tell one another their real names. They have their given names, their work-paper names and "Hey you," as their American supervisors call them. In the telling of their stories, Mercedes and Armando Fernández insisted that their real names be used, to protect their identities. It was their work names they did not want used, names bought in a back alley in Barstow, Tex.

Rarely are the newcomers welcomed with open arms. Long before the Mexicans arrived, Robeson County, one of the poorest in North Carolina, was an uneasy racial mix. In the 1990 census, of the 100,000 people living in Robeson, nearly 40 percent were Lumbee Indian, 35 percent white and 25 percent black. Until a dozen years ago the county schools were de facto segregated, and no person of color held any meaningful county job from sheriff to court clerk to judge.

At one point in 1988, two armed Indian men occupied the local newspaper office, taking hostages and demanding that the sheriff's department be investigated for corruption and its treatment of minorities. A prominent Indian lawyer, Julian Pierce, was killed that same year, and the suspect turned up dead in a broom closet before he could be charged. The hierarchy of power was summed up on a plaque that hangs in the courthouse commemorating the dead of World War I. It lists the veterans by color: "white" on top, "Indian" in the middle and "colored" on the bottom.

That hierarchy mirrors the pecking order at the hog plant. The Lumbees—who have fought their way up in the county apparatus and have built their own construction businesses—are fond of saying they are too smart to work in the factory. And the few who do work there seem to end up with the cleaner jobs.

But as reds and blacks began to make progress in the 1990's—for the first time an Indian sheriff was elected, and a black man is now the public defender—the Latinos began arriving. The United States Census Bureau estimated that 1,000 Latinos were living in Robeson County last year. People only laugh at that number.

"A thousand? Hell, there's more than that in the Wal-Mart on a Saturday afternoon," said Bill Smith, director of county health services. He and other officials guess that there are at least 10,000 Latinos in Robeson, most having arrived in the past three years.

"When they built that factory in Bladen, they promised a trickledown effect," Mr. Smith said. "But the money ain't trickling down this way. Bladen got the money and Robeson got the social problems."

In Robeson there is the strain on public resources. There is the substandard housing. There is the violence. Last year 27 killings were committed in Robeson, mostly in the countryside, giving it a higher murder rate than Detroit or Newark. Three Mexicans were robbed and killed last fall. Latinos have also been the victims of highway stickups.

In the yellow-walled break room at the plant, Mexicans talked among themselves about their three slain men, about the midnight visitors with obscured faces and guns, men who knew that the illegal workers used mattresses rather than banks. Mercedes Fernández, like many Mexicans, would not venture out at night. "Blacks have a problem," she said. "They live in the past. They are angry about slavery, so instead of working, they steal from us."

She and her husband never lingered in the parking lot at shift change. That is when the anger of a long day comes seeping out. Cars get kicked and faces slapped over parking spots or fender benders. The traffic is a serpent. Cars jockey for a spot in line to make the quarter-mile crawl along the plant's one-lane exit road to the highway. Usually no one will let you in. A lot of the scuffling is between black and Mexican.

BLACK AND BLEAK

The meat was backing up on the conveyor and spilling onto the floor. The supervisor climbed down off the scaffolding and chewed out a group of black women. Something about skin being left on the meat. There was a new skinner on the job, and the cutting line was expected to take up his slack. The whole line groaned. First looks flew, then people began hurling slurs at one another in Spanish and English, words only they could hardly hear over the factory's roar. The black women started waving their knives at the Mexicans. The Mexicans waved theirs back. Blades got close. One Mexican spit at the blacks and was fired.

After watching the knife scene, Wade Baker went home and sagged in his recliner. CNN played. Good news on Wall Street, the television said. Wages remained stable. "Since when is the fact that a man doesn't get paid good news?" he asked the TV. The TV told him that money was everywhere—everywhere but here.

Still lean at 51, Mr. Baker has seen life improve since his youth in the Jim Crow South. You can say things. You can ride in a car with a white woman. You can stay in the motels, eat in the restaurants. The black man got off the white man's field.

"Socially, things are much better," Mr. Baker said wearily over the droning television. "But we're going backwards as black people economically. For every one of us doing better, there's two of us doing worse."

His town, Chad Bourne, is a dreary strip of peeling paint and warped porches and houses as run-down as rotting teeth. Young men drift from the cinder-block pool hall to the empty streets and back. In the center of town is a bank, a gas station, a chicken shack and a motel. As you drive out, the lights get dimmer and the homes older until eventually you're in a flat void of tobacco fields.

Mr. Baker was standing on the main street with his grandson Monte watching the Christmas parade march by when a scruffy man approached. It was Mr. Baker's cousin, and he smelled of kerosene and had dust in his hair as if he lived in a vacant building and warmed himself with a portable heater. He asked for $2.

"It's ironic isn't it?" Mr. Baker said as his cousin walked away only eight bits richer. "He was asking me the same thing 10 years ago."

A group of Mexicans stood across the street hanging around the gas station watching them.

"People around here always want to blame the system," he said. "And it is true that the system is antiblack and antipoor. It's true that things are run by the whites. But being angry only means you failed in life. Instead of complaining, you got to work twice as hard and make do."

He stood quietly with his hands in his pockets watching the parade go by. He watched the Mexicans across the street, laughing in their new clothes. Then he said, almost as an afterthought, "There's a day coming soon where the Mexicans are going to catch hell from the blacks, the way the blacks caught it from the whites."

Wade Baker used to work in the post office, until he lost his job over drugs. When he came out of his haze a few years ago, there wasn't much else for him but the plant. He took the job, he said, "because I don't have a 401K." He took it because he had learned from his mother that you don't stand around with your head down and your hand out waiting for another man to drop you a dime.

Evelyn Baker, bent and gray now, grew up a sharecropper, the granddaughter of slaves. She was raised up in a tar-paper shack, picked cotton and hoed tobacco for a white family. She supported her three boys alone by cleaning white people's homes.

In the late 60's something good started happening. There was a labor shortage, just as there is now. The managers at the textile plants started giving machine jobs to black people.

Mrs. Baker was 40 then. "I started at a dollar and 60 cents an hour, and honey, that was a lot of money then," she said.

The work was plentiful through the 70's and 80's, and she was able to save money and add on to her home. By the early 90's the textile factories started moving away, to Mexico. Robeson County has lost about a quarter of its jobs since that time.

Unemployment in Robeson hovers around 8 percent, twice the national average. In neighboring Columbus County it is 10.8 percent. In Bladen County it is 5 percent, and Bladen has the pork factory.

Still, Mr. Baker believes that people who want to work can find work. As far as he's concerned, there are too many shiftless young men who ought to be working, even if it's in the pork plant. His son-in-law once worked there, quit and now hangs around the gas station where other young men sell dope.

The son-in-law came over one day last fall and threatened to cause trouble if the Bakers didn't let him borrow the car. This could have turned messy; the 71-year-old Mrs. Baker keeps a .38 tucked in her bosom.

When Wade Baker got home from the plant and heard from his mother what had happened, he took up his pistol and went down to the corner, looking for his son-in-law. He chased a couple of the young men around the dark dusty lot, waving the gun. "Hold still so I can shoot one of you!" he recalled having bellowed. "That would make the world a better place!"

He scattered the men without firing. Later, sitting in his car with his pistol on the seat and his hands between his knees, he said, staring into the night: "There's got to be more than this. White people drive by and look at this and laugh."

LIVING IT, HATING IT

Billy Harwood had been working at the plant 10 days when he was released from the Robeson County Correctional Facility. He stood at the prison gates in his work clothes with his belongings in a plastic bag, waiting. A friend dropped him at the Salvation Army shelter, but he decided it was too much like prison. Full of black people. No leaving after 10 P.M. No smoking indoors. "What you doing here, white boy?" they asked him.

He fumbled with a cigarette outside the shelter. He wanted to quit the plant. The work stinks, he said, "but at least I ain't a nigger. I'll find other work soon. I'm a white man." He had hopes of landing a roofing job through a friend. The way he saw it, white society looks out for itself.

On the cut line he worked slowly and allowed Mercedes Fernández and the others to pick up his slack. He would cut only the left shoulders; it was easier on his hands. Sometimes it would be three minutes before a left shoulder came down the line. When he did cut, he didn't clean the bone; he left chunks of meat on it.

Mrs. Fernández was disappointed by her first experience with a white person. After a week she tried to avoid standing by Billy Harwood. She decided it wasn't just the blacks who were lazy, she said.

Even so, the supervisor came by one morning, took a look at one of Mr. Harwood's badly cut shoulders and threw it at Mrs. Fernández, blaming her.

He said obscene things about her family. She didn't understand exactly what he said, but it scared her. She couldn't wipe the tears from her eyes because her gloves were covered with greasy shreds of swine. The other cutters kept their heads down, embarrassed.

Her life was falling apart. She and her husband both worked the cut floor. They never saw their daughter. They were 26 but rarely made love anymore. All they wanted was to save enough money to put plumbing in their house in Mexico and start a business there. They come from the town of Tehuacán, in a rural area about 150 miles southeast of Mexico City. His mother owns a bar there and a home but gives nothing to them. Mother must look out for her old age.

"We came here to work so we have a chance to grow old in Mexico," Mrs. Fernández said one evening while cooking pork and potatoes. Now they were into a smuggler for thousands. Her hands swelled into claws in the evenings and stung while she worked. She felt trapped. But she kept at it for the money, for the $9.60 an hour. The smuggler still had to be paid.

They explained their story this way: The coyote drove her and her family from Barstow a year ago and left them in Robeson. They knew no one. They did not even know they were in the state of North Carolina. They found shelter in a trailer park that had once been exclusively black but was rapidly filling with Mexicans. There was a lot of drug dealing there and a lot of tension. One evening, Mr. Fernández said, he asked a black neighbor to move his business inside and the man pulled a pistol on him.

"I hate the blacks," Mr. Fernández said in Spanish, sitting in the break room not 10 feet from Mr. Baker and his black friends. Mr. Harwood was sitting two tables away with the whites and Indians.

After the gun incident, Mr. Fernández packed up his family and moved out into the country, to a prefabricated number sitting on a brick foundation off in the woods alone. Their only contact with people is through the satellite dish. Except for the coyote. The coyote knows where they live and comes for his money every other month.

Their 5-year-old daughter has no playmates in the back country and few at school. That is the way her parents want it. "We don't want her to be American," her mother said.

'WE NEED A UNION'

The steel bars holding a row of hogs gave way as a woman stood below them. Hog after hog fell around her with a sickening thud, knocking her senseless, the connecting bars barely missing her face. As co-workers rushed to help the woman, the supervisor spun his hands in the air, a signal to keep working. Wade Baker saw this and shook his head in disgust. Nothing stops the disassembly lines.

"We need a union," he said later in the break room. It was payday and he stared at his check: $288. He spoke softly to the black workers sitting near him. Everyone is convinced that talk of a union will get you fired. After two years at the factory, Mr. Baker makes slightly more than $9 an hour toting meat away from the cut line, slightly less than $20,000 a year, 45 cents an hour less than Mrs. Fernández.

"I don't want to get racial about the Mexicans," he whispered to the black workers. "But they're dragging down the pay. It's pure economics. They say Americans don't want to do the job. That ain't exactly true. We don't want to do it for $8. Pay $15 and we'll do it."

These men knew that in the late 70's, when the meatpacking industry was centered in northern cities like Chicago and Omaha, people had a union getting them $18 an hour. But by the mid-80's, to cut costs, many of the packing houses had moved to small towns where they could pay a lower, nonunion wage.

The black men sitting around the table also felt sure that the Mexicans pay almost nothing in income tax, claiming 8, 9, even 10 exemptions. The men believed that the illegal workers should be rooted out of the factory. "It's all about money," Mr. Baker said.

His co-workers shook their heads. "A plantation with a roof on it," one said.

For their part, many of the Mexicans in Tar Heel fear that union would place their illegal status under scrutiny and force them out. The United Food and Commercial Workers Union last tried organizing the plant in 1997, but the idea was voted down nearly two to one.

One reason Americans refused to vote for the union was because it refuses to take a stand on illegal laborers. Another reason was the intimidation. When workers arrived at the plant the morning of the vote, they were met by Bladen County deputy sheriffs in riot gear. "Nigger Lover" had been scrawled on the union trailer.

Five years ago the work force at the plant was 50 percent black, 20 percent white and Indian, and 30 percent Latino, according to union statistics. Company officials say those numbers are about the same today. But from inside the plant, the breakdown appears to be more like 60 percent Latino, 30 percent black, 10 percent white and red.

Sherri Buffkin, a white woman and the former director of purchasing who testified before the National Labor Relations Board in an unfair-labor-practice suit brought by the union in 1998, said in an interview that the company assigns workers by race. She also said that management had kept lists of union sympathizers during the '97 election, firing blacks and replacing them with Latinos. "I know because I fired at least 15 of them myself," she said.

The company denies those accusations. Michael H. Cole, a lawyer for Smithfield who would respond to questions about the company's labor practices only in writing, said that jobs at the Tar Heel plant were awarded through a bidding process and not assigned by race. The company also denies ever having kept lists of union sympathizers or singled out blacks to be fired.

The hog business is important to North Carolina. It is a multibillion-dollar-a-year industry in the state, with nearly two pigs for every one of its 7.5 million people. And Smithfield Foods, a publicly traded company in Smithfield, Va., has become the No. 1 producer and processor of pork in the world. It slaughters more than 20 percent of the nation's swine, more than 19 million animals a year.

The company, which has acquired a network of factory farms and slaughterhouses, worries federal agriculture officials and legislators, who see it siphoning business from smaller farmers. And environmentalists contend that Smithfield's operations contaminate local water supplies. (The Environmental Protection Agency fined the company $12.6 million in 1996 after its processing plants in Virginia discharged pollutants into the Pagan River.) The chairman and chief executive, Joseph W. Lutter III, declined to be interviewed.

Smithfield's employment practices have not been so closely scrutinized. And so every year, more Mexicans get hired. "An illegal alien isn't going to complain all that much," said Ed Tomlinson, acting supervisor of the Immigration and Naturalization Service Bureau in Charlotte.

But the company says it does not knowingly hire illegal aliens. Smithfield's lawyer, Mr. Cole, said all new employees must present papers showing that they can legally work in the United States. "If any employee's documentation appears to be genuine and to belong to the person presenting it," he said in his written response, "Smithfield is required by law to take it at face value."

The naturalization service—which has only 18 agents in North Carolina—has not investigated Smithfield because no one has filed a complaint, Mr. Tomlinson said. "There are more jobs than people," he said, "and a lot of Americans will do the dirty work for a while and then return to their couches and eat bonbons and watch Oprah."

NOT FIT FOR A CONVICT

When Billy Harwood was in solitary confinement, he liked a book to get him through. A guard would come around with a cartful. But when the prisoner asked for a new book, the guard, before handing it to him, liked to tear out the last 50 pages. The guard was a real funny guy.

"I got good at making up my own endings," Billy Harwood said during a break. "And *my* book don't end standing here. I ought to be on that roof any day now."

But a few days later, he found out that the white contractor he was counting on already had a full roofing crew. They were Mexicans who were working for less than he was making at the plant.

During his third week cutting hogs, he got a new supervisor—a black woman. Right away she didn't like his work ethic. He went too slow. He cut out to the bathroom too much.

"Got a bladder infection?" she asked, standing in his spot when he returned. She forbade him to use the toilet.

He boiled. Mercedes Fernández kept her head down. She was certain of it, she said: he was the laziest man she had ever met. She stood next to a black man now, a prisoner from the north. They called him K. T. and he was nice to her. He tried Spanish, and he worked hard.

When the paychecks were brought around at lunch time on Friday, Billy Harwood got paid for five hours less than everyone else, even though everyone punched out on the same clock. The supervisor had docked him.

The prisoners mocked him. "You might be white," K. T. said, "but you came in wearing prison greens and that makes you good as a nigger."

The ending wasn't turning out the way Billy Harwood had written it: no place to live and a job not fit for a donkey. He quit and took the Greyhound back to his parents' trailer in the hills.

When Mrs. Fernández came to work the next day, a Mexican guy going by the name of Alfredo was standing in Billy Harwood's spot.

The Story of a Garment Worker

Lisa Liu

Lisa Liu is a garment worker in Oakland, California. She came to the United States from China more than a decade ago but found that while the United States has fewer restrictions on workers, in some ways life as a worker was much less secure than she had expected. She became active in efforts to tell other San Francisco-area garment workers about their rights through an Oakland-based community organization, Asian Immigrant Women Advocates. Compare this essay with the prior essays in this work section. How

SOURCE: David Bacon, "The Story of a Garment Worker," *Dollars and Sense,* Number 231 (September/October 2000), pp. 11–12. Used by permission of the author.

does the garment industry compare with food processing? What can the public do to help stop the kinds of exploitation found in this selection?

I'm a seamstress in a factory with twelve other people. We sew children's clothes—shirts and dresses. I've worked in the garment industry here for twelve years, and at the factory where I am now for over a year.

In our factory we have to work ten hours a day, six to seven days a week. The contractor doesn't pay us any benefits—no health insurance or vacations. While we get a half-hour for lunch, there are no other paid breaks in our shift.

We get paid by the piece, and count up the pieces to see what we make. If we work faster we get paid more. But if the work is difficult and the manufacturer gives the contractor a low price, then what we get drops so low that maybe we'll get forty dollars a day. The government says the minimum wage is $5.75, but I don't think that by the piece we can reach $5.75 an hour a lot of the time.

When we hurt from the work we often just feel it's because of our age. People don't know that over the years their working posture can cause lots of pain. We just take it for granted, and in any case there's no insurance to pay for anything different. We just wait for the pain to go away.

That's why we organize the women together and have them speak out their problems at each of the garment shops. If we stop being silent about these things, we can demand justice. We can get paid hourly and bring better working conditions to the workers.

Our idea is to tell them how to fight for their rights and explain what rights they have. Everyone should know more about the laws. We let them know about the minimum wage and that there should be breaks after four hours of work. We organize classes to teach women that we can be hurt from work. And we've opened up a worker's clinic to provide medical treatment and diagnosis. We do this work with the help of Asian Immigrant Women Advocates here in Chinatown.

We can't actually speak to the manufacturers whose clothes we're sewing because they don't come down to the shops to listen to the workers. So when we have a problem it's difficult to bring it to them. Still, we've had campaigns where we got the manufacturer to pay back-wages to the workers after the contractor closed without paying them. We got a hotline then, for workers to complain directly to the manufacturers. That solved some problems. The fire doors in those shops aren't blocked anymore, and the hygiene is better.

But it's not easy for women in our situation, and many are scared. Because they only work in the Chinese community, they're afraid their names will become known to the community and the bosses will not hire them. That's why we try to do things together. There's really no other place for us to go. Most of us don't have the training or the skills to work in other industries. We mostly speak just one language, usually Cantonese, and often just the dialect Toishanese.

When I first came to the United States I needed a lot of time to work to stabilize myself. So after seven years that's why I'm only now having my first baby. We don't have any health insurance and we have to pay the bill out of our own pockets. Health insurance is very expensive in the United States. We can't afford it. In the garment industry here they do not have health insurance for the workers.

Before I came here, my experience in China was that life was very strict. I heard that in America you have a lot of freedom, and I wanted to breathe the air of that freedom. But when I came here

I realized the reality was very different from what I had been dreaming, because my idea of freedom was very abstract. I thought that freedom was being able to choose the place where you work. If you don't like one place, you can go work in another. In China you cannot do this. When you get assigned to a post, you have to work at that post.

Since I've come to the United States, I feel like I cannot get into the mainstream. There's a gap, like I don't know the background of American history and the laws. And I don't speak English. So I can only live within Chinatown and the Chinese community and feel scared. I cannot find a good job, so I have to work the low-income work. So I learned to compare life here and in China in a different way.

Many people say life here is very free. But for us, it's a lot of pressure. You have to pay rent, living costs so much money, you have all kinds of insurance—car insurance, health insurance, life insurance—that you can't afford. With all that kind of pressure, sometimes I feel I cannot breathe.

Everywhere you go you just find low pay. All the shops pay by the piece, and they have very strict rules. You cannot go to the bathroom unless it's lunch time. Some places they put up a sign that says, "Don't talk while you work." You're not allowed to listen to the radio.

Wherever you go, in all the garment factories, the conditions and the prices are almost the same. The boss says, "I cannot raise the price for you and if you complain any more, then just take a break tomorrow—don't come to work." So even though I can go from one job to another, where's the freedom?

REFLECTION QUESTIONS FOR PART IV

The readings in this section focus on the effects that institutions have on individual lives. Reflect on the following general questions from Part IV:

1. What effect does housing (affordability, availability, quality, neighborhood) have on a person's life experience?
2. How does the welfare system affect people receiving welfare? How have changes in the system affected individuals? How do people outside the system view people on welfare?
3. How does the health-care system affect an individual's life chances and quality of life? How can this system be improved?
4. If work is central to a person's identity, what does it mean for those who do society's "dirty work"? What types of exploitation are evident in the selections on work?
5. Compare/contrast the readings in Chapters 6 through 9. What common threads do you see in their experiences? How do race, class, gender, age, and location affect people's experiences with society's institutions?

6. Analyze each article carefully and group them by individual/cultural and structural theories of poverty. Why do you think one theory or another is best reflected in each reading? Explain your reasons.
7. Select one particular essay, such as Luis Rodriguez' "Always Running," and analyze it first *without* using a sociological imagination, and then analyze it again *with* a sociological imagination.

PART V

Individual and Collective Agency and Empowerment

The curse of poverty has no justification in our age. It is socially as cruel and blind as the practice of cannibalism at the dawn of civilization. . . . The time has come for us to civilize ourselves by the total, direct, and immediate abolition of poverty.

MARTIN LUTHER KING, JR. (1967)

10

Changes from the Bottom Up

The discipline of sociology focuses on the social context and the social forces that strongly affect human behavior, but while society and its structures are powerful, the members of society are not totally controlled. We are not passive actors. We can take control of the conditions of our lives. Human beings cope with, adapt to, and change social structures to meet their own needs. Individuals acting alone or with others can shape, resist, challenge, and sometimes change the social institutions that impinge upon them. These actions constitute **human agency.** The essence of agency is that individuals, through collective action, are capable of changing the structure of society and even the course of history. But while agency is important, we should not minimize the power of the structures that subordinate people, making change difficult or, at times, impossible.

At the individual level, the poor and the near poor can use various strategies to supplement their meager resources. Single mothers, for example, can do income-producing work such as house cleaning, laundry, mending, childcare, and selling homemade items (Harris, 1996; Schein, 1995), as well as giving and receiving help from family, friends, neighbors, boyfriends, and absent fathers. Those strategies are ways to cope with and adapt to the existing structural limitations. The focus here, however, is on joining with others to change social structures—to make the structures more equitable.

---

*This introduction is taken largely from Eitzen and Baca Zinn, 2001:525–531.

Individuals seeking change have little effect. If, however, they join with others who share their convictions and goals, and if they organize and map out a plan of action, they can make a difference (Kammeyer, Ritzer, and Yetman, 1997:632–633). In doing so they create a social movement. A **social movement** is an issue-oriented group specifically organized to promote or resist change. Such movements arise when people are sufficiently discontented that they are willing to work for a better system.

At various times and places, the poor and the near-poor have organized to change oppressive situations. Janitors, hotel maids, garbage collectors, women on welfare, farmworkers, and others have sometimes succeeded in making the powerful meet their demands. Historically, the most successful social movements by the poor and powerless were the Montgomery bus boycott led by Martin Luther King, Jr. in the 1950s and the organization of farmworkers by Cesar Chavez in the early 1960s. King and his organization of African Americans ended segregated busing in the South with a Supreme Court decision and led to many more civil rights victories (Branch, 1988). The campaigns by Chavez and his farmworkers, through grape boycotts and other tactics, led to a 1966 agreement with the farm owners, requiring the owners to provide clean drinking water, rest periods, and other benefits for the workers, and to the 1975 passage of the California Agricultural Labor Relations Act—the first law in the nation to guarantee farmworkers the right to form unions and to bargain collectively (Griswold del Castillo, 1995).

Recently, campaigns by the poor and nonpoor alike have resulted in the institution of living-wage ordinances in more than 60 cities (including Baltimore, New York City, Boston, and Los Angeles) to help lift low-paid workers out of poverty. These efforts continue, as do collective efforts to eliminate sweatshops and to remove farmworkers from exposure to pesticides.

The essays in this chapter provide case studies of poor individuals and groups of poor people who have sought and achieved change.

NOTES AND SUGGESTIONS FOR FURTHER READING

Branch, Taylor (1988). *Parting the Waters: America in the King Years 1954–63.* New York: Simon and Schuster Touchstone Books.

Clark, Christine, and James O'Donnell (1999). *Becoming and Unbecoming White: Owning and Disowning a Racial Identity.* Westport, CT: Bergin and Garvey.

Eitzen, D. Stanley, and Maxine Baca Zinn (2001). *In Conflict and Order: Understanding Society,* 9th ed. Boston: Allyn and Bacon.

Griswold del Castillo, Richard (1995). *Cesar Chavez: A Triumph of Spirit.* Norman, OK: University of Oklahoma Press.

Harris, Kathleen Mullan (1996). "The Reforms Will Hurt, Not Help Poor Women and Children." *The Chronicle of Higher Education* (October 4):37.

Jennings, James (1994). *Blacks, Latinos, and Asians in Urban America: Status and Prospects for Politics and Activism.* Westport, CT: Praeger.

Jennings, James (ed.), (1997). *Race and Politics: New Challenges and Responses to Black Activism*. London: Verso.

Kammeyer, Kenneth C. W., George Ritzer, and Norman R. Yetman (1997). *Sociology*, 7th ed. Boston: Allyn and Bacon.

Kern, Jen (2001). "Working for a Living Wage." *Multinational Monitor* 22 (January/February):14–16.

Lofland, John (1996). *Social Movement Organizations: Guide to Research on Insurgent Realities*. New York: Aldine de Gruyter.

Morris, Aldon (1984). *The Origins of the Civil Rights Movement*. New York: Free Press.

Piven, Frances Fox, and Richard A. Cloward (1977). *Poor People's Movements: Why They Succeed, How They Fail*. New York: Pantheon.

Pope, Jackie (1999). "Women in the Welfare Rights Struggle." Pp. 287–304 in *A New Introduction to Poverty: The Role of Race, Power, and Politics,* Louis Kushnick and James Jennings (eds.). New York: New York University Press.

Schein, Virginia E. (1995). *Working from the Margins: Voices of Mothers in Poverty*. Ithaca, NY: Cornell University Press.

Wright, Carter (2001). "A Clean Sweep: Justice for Janitors." *Multinational Monitor* 22 (January/February):12–14.

Wright, Talmadge (1997). *Out of Place: Homeless Mobilizations, Subcities, and Contested Landscapes*. Albany, NY: State University of New York Press.

Wardell: Resident of Rockwell Gardens

Susan J. Popkin, Victoria E. Gwiasda, Lynn M. Olson, Dennis P. Rosenbaum, and Larry Buron

In a previous selection in Chapter 5, we heard from Dawn, a resident of Rockwell Gardens, a public-housing development in Chicago. Here another resident of that community, unlike Dawn, is taking steps to do something about the problems in his community. Compare/contrast this selection with Dawn's words in Chapter 5. What steps has Wardell taken to improve his community? What problems has he faced in his personal struggle? What can we learn from Wardell's experience, especially in comparison with Dawn's?

WARDELL'S STORY

While most of Rockwell was a place of extreme hardship and social decay, one building in the development was set apart from the rest of the community. In the early 1990s, a group of residents, led by a charismatic man in his early fifties named Wardell Yotaghan, started a resident management corporation in their building.

Wardell's version of life in Rockwell is very different from Dawn's. Unlike many Rockwell residents, he had not spent his whole life in CHA housing; he had moved into the development a decade ago when he married the mother of his three youngest children. He had, however, spent his entire life living on the west side of Chicago and witnessed the devastation that drugs have wrought on his community.

Wardell viewed himself as having been saved from the streets by athletic programs; he said he was inspired by Muhammad Ali's example. He said that most members of his family *have* made it—his mother and sisters worked, and

SOURCE: Susan J. Popkin, Victoria E. Gwiasda, Lynn M. Olson, Dennis P. Rosenbaum, and Larry Buron, *The Hidden War: Crime and the Tragedy of Public Housing in Chicago* (Piscataway, NJ: Rutgers University Press, 2000), pp. 43–45. Used by permission of the publisher.

one sister went to college and has her own business. Wardell finished high school and attended college off and on over the years. He worked most of his life as a security guard, but he gave that up when he became a resident activist and president of his building in the early 1990s.

Despite these successes, Wardell had been affected by the same problems that Dawn describes. Over the past two decades, he had lost several friends and relatives to drugs:

> Well, the truth is, most of the people I grew up with, in the area I grew up . . . most of them are dead. . . . It [heroin] took a big toll in the area where I lived. Me and a guy I was talking to a few days ago, he was younger than me, but we come from the same area, we were talking about all the people that were dead that we used to play ball with. When we were young, there were things for us to do. . . . There was basketball, softball, swimming, and those were things that we done. And the guys that we used to do that with, they're all gone.

This destruction had driven Wardell to action; he said his mission in life was to "clean up the mess" that drugs and drug trafficking have created in his co munity. He had become a full-time community activist: advocating for Rockwell residents, bringing programs and services to his building, and, more recently, fighting to preserve low-income housing in Chicago while the CHA plans to demolish many of its family developments.

Like Dawn, Wardell was particularly concerned about the children of Rockwell. He saw children suffering because their mothers, addicted to drugs, were exposing them to violence: "it ain't no secret that a lot of the people in the building . . . are addicted to drugs. So they go out and these people [the dealers] give them this credit and they don't pay, and here they [the dealers] come up to the apartment and I'm not as much concerned about what they do to that individual as I am about the terror that they bring on the children that she have in the apartment."

Wardell was also concerned about the gangs in Rockwell, but unlike many residents, he was not intimidated by them. In his view, the gangs mostly shot at each other, not at other residents. Wardell said that usually the "gangbangers" warn other residents to get out of the way. He was only upset by the shooting when it put school children in danger:

> They [kids in his building] usually play down in the playground. And, usually if there's gonna be some shooting, the guys that either gonna get shot at or do the shooting tell the kids to go upstairs. . . . So, even when they were shooting on a regular basis . . . we knew what time to go down. The only part that we really was concerned about is that they shoot right at the time the school is taking in and right at the time the school is letting out. I never could understand that.

Wardell had worked with other residents to make his building an oasis from the drugs and violence that pervade Rockwell. The tenants in his building had organized a resident management corporation in 1991 and tried to create a healthy community. According to Wardell, they were helped by the fact that their building sits on the northeast edge of the development, cut off from the other buildings in Rockwell by a major street. Further, the building was being remodeled; because half the units were empty for more than five years, it was easier to form a cohesive group. Despite these advantages, Wardell said the building had had frequent problems with drug dealers; attempts to reduce the problems with crime and drug trafficking bring with them potential risks. "Certainly it takes the tenants' participation to get rid of crime. But if I'm the police, and you call the police, and I come and I say, 'Where's Miss So and So?' Well, that turns people off from calling the police and giving them information. . . . Some tenants will [still] do that. But to me they risk their lives and the lives of their family because these people [drug dealers] are making large sums of money."

Yet, despite his very real concerns about retaliation, Wardell had called the police for help with specific problems, complained to the CHA, and worked with other tenants to push the dealers out of his building. Gradually, he felt this had had an effect, although it remained a constant struggle. "We went to the people that were selling it in our building and said, 'Look, we're trying to do something here. We want you to don't sell drugs here.' And they sort of didn't pay us much attention, but they started looking to see were we really doing something, and when they saw we were really trying to do something, they moved it out."

Surviving Chicago's Toxic Doughnut

Hazel Johnson

Hazel Johnson, a 64-year-old African American woman, talks here about her fight against the toxic environment surrounding public housing in Chicago. This selection is an example of environmental racism/classism (the concentration of poor and racial minorities near hazardous and toxic environments). Can you think of other examples of environmental racism in different areas of the country? What does her organization (PCR) do to promote social change, and how does this demonstrate collective empowerment?

I am a mother of seven children and nine grandchildren. I have been a resident of the Altgeld Gardens community for 37 years. Altgeld Gardens is a public housing development located on the Southeast side of Chicago. I became a community representative, taking children on field trips to the amusement park and other places, during the summer months. I volunteered at the local parent school council and was elected to the Altgeld Local Advisory Council.

SOURCE: Hazel Johnson, "Surviving Chicago's 'Toxic Doughnut,'" Environment Justice Resource Center (March 24, 2001), http://www.ejrc.cau.edu/voices from the grassroots.htm. Reprinted by permission of Robert Bullard.

I started People for Community Recovery (PCR) in 1979. It was started as a group of women organizing on environmentally-related health problems in this community. PCR was incorporated October 25, 1982. Our organization is one of the first African American grassroots community-based environmental organizations in the Midwest. Our mission is to address multiple exposures to harmful toxins and pollutants surrounding public housing. For the past 20 years, I have been active in environmental issues in my community and other communities of color around the country. I got involved in environmental issues while watching the news and learned that the Southeast side of Chicago had the highest incidence of cancer of any community in the city.

Later, I connected with the city and state health departments. These agencies mailed me many reports on environmental problems in Southeast Chicago. PCR conducted its own land use survey of the neighborhood. We began knocking on my neighbors' doors asking them to fill out the health survey. We learned that people were suffering with severe health problems, including asthma, cancer, skin rashes, kidney and liver problems. To no one's surprise, we found alarming patterns. The Southside neighborhood, Altgeld Gardens in particular, was surrounded by all kinds of polluting industries, landfills, incinerators, smelters, steel mills, chemical companies, paint manufacturing plants, and a municipal sewage treatment facility. My neighborhood is also surrounded by more than 50 abandoned toxic waste dumps. We live in a "toxic doughnut."

Despite poor environmental conditions in our community, this did not discourage our group from wanting to learn more about the environmental conditions and the possible impact on residents' health. PCR began organizing residents to get the neighborhood cleaned up and treated fairly. For the past decade, we pressured corporate polluters, the city, and state officials to make them aware of their negligence and make them accountable. It has not been easy going up against the giant corporations, but we are fighting a life-and-death struggle. Through perseverance and dedication, we have successfully brought the needed attention to the environmental issues in Southeast Chicago. We have to fight for our children. We have educated ourselves on environmental issues and the health threats from nearby polluting industry. We have not waited for government to come in and determine the "cause" of our illnesses. We may not have Ph.D. degrees, but we are the 147 "experts" on our community.

In 1992, PCR undertook its own health survey of 825 Altgeld Gardens' residents. We were joined by volunteers from the University of Illinois School of Public Health (designed the survey instrument), Clareitian Medical Clinic (conducted training of interviewers), and St. James Hospital (designed the graphs). Their goal was to follow up on the long-standing anecdotal evidence of health problems. The results of the survey were no surprise. In addition to heightened risks of troubled pregnancies, the survey revealed a high incidence of chronic pulmonary disease, which includes emphysema and chronic bronchitis. Thirty-two percent of men and 20 percent of the women surveyed had asthma. Sixty-eight percent of those surveyed indicated that they experienced health problems that disappeared when they left Altgeld Gardens. More than 37 percent of the respondents cited noxious odor when asked to comment generally on their most common complaint.

The environmental justice work that we started in Chicago has allowed me to testify before Congress and meet two presidents of the United States. Our group has sponsored "toxic tours" of the community with dignitaries from around the world. We have hosted two environmental conferences. We are often asked to speak at universities and colleges, at workshops and training programs about urban environmental pollution and racism. Our environmental justice work has kept us busy. More importantly, it has paid off.

PCR's organizing efforts persuaded the Chicago Housing Authority and Chicago Board of Education to remove asbestos from the homes and schools in Altgeld Gardens. We assisted elderly tax-paying residents of Maryland Manor, another Chicago housing development, in getting water and sewage services. Our group also shut down a nearby hazardous waste incinerator and fought to get a comprehensive health clinic in the southside neighborhood.

PCR along with other people of color grassroots groups took their struggle to the Rio Earth Summit where they were joined in solidarity with other brothers and sisters around the world who are experiencing similar environmental and economic injustices. It did not take me long to realize that the environmental, economic, and health problems in the favelas of Rio de Janeiro looked a lot like the problems in my Southside Chicago neighborhood.

Our organization is growing and maturing, and we are still learning. In 1992, PCR was a recipient of the President's Environmental and Conservation Challenge medal, the nation's highest environmental award. That was a great honor. However, the biggest award and honor I could get from government officials right now is for them to "do the right thing," by making the polluters clean up their act on the Southside.

PCR has formed alliances and coalitions with national environmental and civil rights organizations as well as other local grassroots environmental groups. We see ourselves as an integral part of the environmental and economic justice movement. Poor people and people of color must empower themselves to become politically active. Everything is political. We must learn how to fight for environmental justice at home and abroad. No one will save our communities but us.

Welfare Rights Organizing Saved My Life

Dottie Stevens

Dottie Stevens, former welfare recipient and battered mother, talks about becoming involved in welfare rights and what that did for her life. What words or phrases that are used throughout this essay can be associated with empowerment? How would individual/cultural and structural theorists analyze her situation and actions? What can be learned from her experiences?

I grew up in east Boston, Massachusetts, a neighborhood with a large Italian population (mostly immigrants). For me, they represented large families, strange accents, and wonderful smells from delicious mysterious foods. Wine was on the table for all meals, and even the young children were encouraged to sip it from fancy glasses. My street was a mixture of nationalities. We were also Irish, French, English, Native American, and Chinese, as well as many other mixtures we couldn't figure out. My own background was Irish and German.

SOURCE: Dottie Stevens, "Welfare Rights Organizing Saved My Life," *For Crying Out Loud: Women's Poverty in the United States,* Diane Dujon and Ann Withorn (eds.), (Boston: South End Press, 1996), pp. 313–316. Reprinted by permission of the publisher.

On the other side of my street was a housing project that went on for four blocks. It became my playground and community. My own home was a two-family house. Nana lived on the first floor and my mother, two brothers, two sisters and I occupied the second floor. Mom was divorced and had to apply for welfare to keep us together. I don't remember seeing Mom much after she began to work under the table, because the welfare's cash benefits wouldn't cover all the bills.

I found that if I wanted attention or something to eat, I could go to any of my friends' homes on the street or in the projects and blend in with their families, and feel I belonged. I liked to stay at my friend Angela's house the best. She had a mother *and* a father at home and they always had fresh, hot Italian bread to share with me. Angela had one brother, and many aunts and cousins, but what I was most impressed with were her four handsome uncles! They all communicated loudly, bellowing passionately to everyone in their space, arms and hands slicing through the air. To me, that noise and confusion represented what family was *supposed* to be.

I am a product of the system, so to speak. I was a child of five years old in a single-female-headed household with four siblings, and we were on welfare. Memories of everyday living were not too bad. We were used to having no heat until Thanksgiving, and we never had hot water in the summertime. We lived in our own home and envied the families living across the street from us in the projects. They *always* seemed to have heat and hot water all year round. Their electricity *never* got shut off. They received repairs if anything broke down and had their apartments painted every other year automatically. It seemed to me that living in the projects was *luxury*!

My mother was (and is) a proud woman and a very hard worker. As an adopted, only child of a doting, elderly, immigrant couple, she was ill-equipped to cope with five children in poverty and an ex-husband who never paid her any child support. But she did the best she could.

Mother did not like her social worker, Mrs. Brown, who always made her feel like a loser. She would come to our house unannounced and ask to look in closets and cabinets as if my Mother was hiding something or someone. To this day, I can see the tension and feel the anxiety this worker provoked in my mother's life.

We did not have a bad life; we did not really know we were poor. We did not own a car; we had no phone. But Mother managed, somehow, to get us all into parochial school. She always told us education was the key to success, although she had had very little education herself. My mother's influence lasted only until the eighth grade. After all, I was a teenager and my mother was the *last* person I listened to!

At 16 years old (after another fight with my mother) I eloped, thinking it was the way to a better life. (Wasn't the point to "at least get him to *marry* you"?) A son was born of this union, the only good part about it. Although I was madly in love, after my first child was born, reality set in and I knew my marriage wouldn't last. Two years later, we parted and my world fell apart. I was only 18 years old!

I loved my baby son, but alone with no income, I did not think I would be able to keep us together. So I went to Mom's house for help.

I divorced at 20, worked for several years at factory work and waitressing jobs, went to night school for a while, then married again. This marriage lasted six years. I had two more children and, when the fighting got real bad, this husband just disappeared, and we have not seen him since. At 28, I was divorced again!

I had just about given up on the love-and-marriage bit; I went back to night school and was set to stay a single parent, when husband #3 showed up on the scene. He was kind, affectionate, and took an interest in me and my children. I fell passionately in love and thought this was meant to be!

I had married this last time out of my culture, into one I could not understand. Things that seemed odd and uncommon to me, I chalked up as different ways, traditions, customs, and language. We were different! I went along with it: different dress, different conversations, recreations and working conditions. I just wanted to be a good wife and mother, to take care of the home, go to the PTA meetings, bake, shop, keep myself attractive, and be there for my man. But through day-to-day experiences, I eventually realized what was happening, and this was not the reality I had expected for my family!

What had happened to love? This was not what I thought it would be. Yes, he professed to love me and the children and showed compassion and kindness. Yes, he vowed to love and care for us forever. Yes, he could be sweet and gentle, but look out! *Emotionally, Freddy*

Kruger lived here. I couldn't close my eyes for too long; if I slept, anything could happen in the dark.

I knew I would have to hide from him because he was not about to let us go. I was running, escaping for my life and my sanity! With my children (the baby in my arms and the three others in tow), I fled from an abyss of cruelty, deceit, and oppression that I had never known existed. I had been a romantic fool, brainwashed to believe in love.

I stayed in a battered women's shelter for a month. I finally managed to get him out and get our home back. When I looked toward the future, it seemed desperately bleak and uncertain. My self-esteem was nil and I had no money. My children and I had been physically, mentally, and emotionally battered for a long time and needed many years of therapy and counseling to be able to deal with it; and after two decades, we are still haunted by the abuse. When my youngest child went to school, I knew I needed to do something to better myself economically. We had been subsisting on welfare benefits that could not decently support a family, so I thought I should go back to school and earn a degree that would enable me to earn a breadwinner's wage. I had never gone to high school and was told I needed a GED to get admitted to college at U-Mass/ Boston, College of Public and Community Service (CPCS). When I was eventually admitted to CPCS, the first thing I did was buy four pairs of knee-socks and I was ready for a new start in my life!

In September 1979, I was finally on my way—to where or what, I had no idea, but I felt it would be O.K. I was getting educated! This college was not what I expected; it was comprised of adult working people and many single mothers on welfare trying to do the same thing as I. I had found my niche and I thrived!

In my second semester, I entered a class called "Basic Organizing." This class was made up of 13 welfare-recipient mothers and a few human service professionals. Ann Withorn was our instructor, and she made me feel as if I had come to the right place. Was I in for a surprise when I found out *my* life experience was really a valuable asset in her class! I had just ended my third marriage; and I was a former abused and battered wife of a substance abuser.

Some of the other student mothers were so smart, I thought I could not keep up, but we learned to become a self-help group and helped each other in many ways to earn credits by working together in groups. And what a dynamic group we were! We actually became a sisterhood—sort of a sorority of poor women—sharing our most intimate secrets, experiences, hopes, dreams, and disappointments. We developed a collective knowledge, which we were able to use to help each other fend off the Welfare Department. We adopted a principle to never go to the Welfare Department alone—we became each other's advocates at fair hearings. We exchanged clothing, shared childcare responsibilities when our childcare arrangements fell through, compiled lists of wealthy suburban communities' trash pick-up days, and organized furniture shopping sprees for the nights before trash collections. Individually, we were weak, but together we were formidable!

The "Basic Organizing" class began organizing on campus and developed into a Recognized Student Organization, which allowed us to obtain office space, a telephone, and a small budget with which to work. We voted on a name and were now called "Advocacy for Resources for Modern Survival" (ARMS). We voted to become a chapter of The Coalition for Basic Human Needs (CBHN), which was the statewide welfare rights organization. Our priority was to advocate for welfare recipients' access to higher education. We began to serve on many advisory committees and on boards of directors of antipoverty agencies that affected our lives, such as daycare centers, health care providers, fuel assistance agencies, tenants' rights organizations, clothing and food distribution centers, etc.

Many of us felt intimidated and unable to even know what was being said at the meetings, but we took notes, came back to ARMS, and discussed all that we had heard and educated each other. This was a wonderful time getting closer to the other women and their children and beginning to trust each other. Working together, we were able to keep ourselves informed on several fronts at once. We were able to monitor proposed changes in the service delivery of a variety of agencies, assess their often conflicting policies, make holistic appraisals of their effects on our lives, develop recommendations, and, most importantly, significantly impact the final decisions. *We were making a difference!*

In this "Basic Organizing" class, the group decided to organize around welfare issues. The Aid to Families with Dependent Children program was very important to all of us because we derived our income and health benefits from it, but it was an inadequate, punitive system that harassed and demeaned us. We all decided to fight back using techniques that would enable us to earn our sheepskins. We started to research poverty issues going all the way back to the Elizabethan Poor Laws, and we found out that poor people *always* had it bad! There were the manors and the serfs, the Pharaohs and the slaves, the Old South and the Blacks, and now we had billionaire politicians and Workfare participants!

Learning the theories, history, and practice of organizing saved my life; actually, it brought me to life! I finally had something worthwhile to live for.

Josefina Flores: A Veteran of the War in the Fields

David Bacon

Josefina Flores is a field organizer for the migrant workers' union, United Farm Workers. At the time of this interview, she was engaged in a campaign to organize the strawberry industry in the face of sometimes violent opposition by the companies involved. Think about the role of unions in the lives of poor, minority laborers. What are the pros and cons of labor unions? What particular problems exist in trying to collectively organize farm laborers?

Josefina Flores, a woman in her sixties, looks a little weather-beaten, from a life spent in the sun. "I was born in Calexico, in the Imperial Valley," she remembers. "I worked in the fields from the time I was 7 years old, in the lettuce, in the beets, the carrots, and finally, the grapes. My family worked the *corrido*—we traveled with the crops, making the same circuit every year."

"In those years, no one ever demanded that the children of farm workers go to school. I just went a few days in one place and a few days in another. I wanted to go to school. I could see the other children playing and learning things. But I couldn't, because of the economic situation of my family. I was the oldest child—I had to help, so I

SOURCE: David Bacon, "Josefina Flores: A Veteran of the War in the Fields," *Dollars and Sense*, Number 231 (September/October 2000), pp. 38–39. Used by permission of the author.

went to work. This was in 1937 and 1938, and the wages were 30¢ an hour. At first I just worked helping my parents. Adding what I did to their work, we all just made a little more."

Flores was born in the United States, and so is a U.S. citizen. "But even so, we were just seen as poor people, as Mexicans," she recalls. "Being poor and being Mexican was just about the same thing."

She got married at 21, and had nine children. Only three are still living, and her husband died when the oldest was 10. "Afterwards, I just went on working," she says. "I brought my kids to the fields, the way my parents brought me. We had a little *corrido*—just from Arvin to Selma and back [two small towns in California's Central Valley]."

Today, Flores is an organizer for the United Farm Workers, a veteran of the labor wars that have swept California fields since the 1960s. Driving through the Pájaro Valley around Watsonville, she stops from time to time to speak with workers at the edge of the strawberry fields. There they work, doubled over, for ten hours and more every day. Most are very young—eighteen or twenty years old.

Flores calls them to come talk with her. She takes down, in a thick, spiral-bound notebook, careful notes about what they say. Every name is written down—what each person says about their wages, the problems of their job, the name of their foreman. She keeps this record partly to help the union organize. But who else is going to write down what happens to these young men and women of the strawberry fields? The spiral-bound notebook is the record that each person was here, that each thought a certain way about their situation. Josefina Flores has become an archivist for the voiceless.

"I never learned to read and write when I was a child—not until much later," she says. "I learned in the union. Organizers would teach me a little here, and a little there, long after I was already grown up."

Flores became a member of the United Farm Workers when the union first began. "In 1965 we began to hear that César Chávez was organizing a union for farm workers. I didn't know what a union was," she explains. "I was working on a ranch near Selma, and there was a strike. I saw that the union was trying to do something good, just trying to get the wages up a little, and get bathrooms in the fields. So I joined."

In 1968, the union struck the grape harvest, as it did every year from 1965 to 1970. That year, Flores went with her daughter to collect food donations at a local restaurant. "While I was inside, a foreman came in and shot me. He shot seven bullets into my body." It took Flores two years to recover. Afterwards, she became a UFW organizer and a leader of the huge 1973 grape strike, when growers tore up their union contracts and brought in the Teamsters. That epic battle was called off after two other workers were shot and killed on the picket line.

When Flores looks at the young immigrants who make up the workforce in the strawberry fields today, she sees workers who have little idea of the sacrifice she and the other union veterans made over the years to build a union. "People come here from far away in Mexico," she explains. "They know very little about their rights as workers. We have to explain the whole history of the farm workers' movement to people who not only weren't here when it happened—many of them hadn't even been born. To them, César Chávez is a boxer." (She refers to Julio César Chávez—the Mexican boxing champion.)

"If they don't come here [to the United States], they don't eat," she adds. "They tell me—'we're not here because we want to, but because we have to.' In the little towns they come from, there's no clinic, and sometimes no school. They can't read and write. And of course the labor contractors threaten them. A lot of workers have no papers, so it's not hard to scare them. That's the biggest problem—for us and for them. They know so little."

Flores looks at the young workers and sees herself as she was when she was young. "They're tired of being poor and ground down," she concludes, "just like I was. And while some are afraid, these young people are still not as accepting of things as farm workers were when I was their age. Their expectations are higher than ours were. That's why I think our movement can help them."

"They want something better. I hope they're willing to fight to get it."

Apologies Don't Help

Milwaukee Welfare Warriors

This letter was written by the Milwaukee Welfare Warriors in the fight for the lives of mothers and children. What does this group mean when it says "apologies don't work"? What apologies are they referring to? This statement was written before the 1996 welfare reform legislation (see Chapter 7). How do you think the Milwaukee Welfare Warriors would respond to this legislation?

Thank you for your help in publishing facts and myths about those of us who receive government child support. May we ask you to go one step further in your support of our families?

Popular lists of "facts" about welfare are defensive: We *only* have two children, *only* stay "on" welfare for two years, *only* receive $370 a month, *only* use up 1 percent of the federal budget, *only* need help temporarily to get us on our feet, would "*work*" if *only* we could afford childcare/health care or could find a job, are *mainly* white adults, not teens, and *mainly* children.

These apologetic "facts" present statistical truths about welfare. However, they omit two profound realities of welfare:

1. AFDC is a public child support program.

2. Most single-mother families on welfare are victims of abuse and/or abandonment.

No other moms are called *dependent* or made to feel like parasitical, apologetic criminals for receiving support for their children. Widowed moms, some divorced moms, married moms all expect and receive support—from both the government (tax deductions, Social Security) and the biological fathers. Neither they nor their children are accused of being social deviants or mentally defective (low self-esteem, etc.) because they receive economic support. Nor are they labeled "recipients"—an insulting, passive, one-dimensional label of the complex being a single mother on welfare is.

And what about those of us with three, four or five children, or those of us who are teen moms? What about the moms who can't both raise kids alone and work full-time? What about the women who aren't white? What about those of us who use the support for far more than two years? Most of all, what

SOURCE: Welfare Warriors publish a 28-page Internatinal Activist Mothers' newsjournal, *Warrior Mothers Voice*. Contact them at 2711 W. Michigan, Milwaukee, WI 53208, 414-342-6662.

about the vast majority of us who will never get our families out of poverty with one woman's salary?

Apologetic statistics are not working to convince Americans that children and mothers have a right to share in our wealth. Apologies are not convincing taxpayers that children need support, even if mom is employed. Apologies are not stopping the violence and terrorism of welfare reform. Apologies are not stopping the government from taking children away from loving homes. Apologies are not stopping the government from giving the majority of our tax money to corporations and the Pentagon. Apologies are not creating living-wage jobs for moms (or dads). Apologies are not helping Americans understand the problem or the solutions. Apologies are not helping single mothers and children retain the strength needed to fight back.

It is time for our allies to do more than apologize for our existence. It is time to stand up for our right to public support for our children and our right to do paid work or get help from a partner without losing that support. It is time for our allies to state loud and clear that they will not tolerate systematic punishing of mothers and children for being victims of abandoning dads and a slave-wage work force.

It is time for our allies to fight for us, not apologize for our existence.

REFLECTION QUESTIONS FOR PART V

After finishing the section on agency, reflect on the following questions from Part V.

1. Think of the terms *agency, empowerment,* and *collective action*. How does each of the readings demonstrate these terms?
2. Which do you think is more empowering—individual action or collective action? Why?
3. Compare/contrast the readings in this section with one another. Which methods are more effective in bringing about change? Why?
4. Think about this final section in relation to previous ones. What lessons can you take from this section regarding housing, welfare, parenting, schooling, and work?
5. Select one particular essay, such as Popkin et al., "Wardell: Resident of Rockwell Gardens," and analyze it, first *without* using a sociological imagination and then analyze it again *with* a sociological imagination.

Afterword

We end this book of voices of the poor with a sermon by a Unitarian Universalist minister who grew up in poverty. Reverend Bumbaugh, with a keen sociological imagination, reflects back on his childhood and then looks at the condition of the poor in contemporary society. He is scandalized and wants all of us to be scandalized by an affluent society that is structured in ways that allow its poor and especially its poor children to suffer. Along with sharing his insights about poverty, he implores us to be morally outraged by a society that turns its collective back on its poor children.

The Forgotten Children of Affluent America

David E. Bumbaugh

As I prepared for this sermon, I frequently heard a voice in my inner ear speaking with great authority. "It's no disgrace to be poor," said my Aunt Martha, talking like an adult to two small boys who desperately wanted something she could not give them. "Why," we had asked, "do we have to be poor?" "It's no disgrace to be poor," she repeated, more to herself than to us, "it's just damned inconvenient!" It was the kind of comment I would hear her make many times as we were growing up. As she sat at her old treadle-powered sewing machine, manufacturing shirts for us and blouses and dresses for herself out of old flour sacks, she would mutter with a sigh, "It's no disgrace to be poor. We should be grateful for what we have." As we stood together in the kitchen, working over a pile of spotty green beans or tomatoes well past their prime or corn left to grow tough on the cob, culling the usable from the useless and preparing it for canning so there would be food for the winter, she would murmur, more to herself than to me, "It's no disgrace to be poor, but does it have to be so tough?"

As she conducted her frequent war against the vermin which inhabited the walls of the old apartment building in which we lived, sprinkling a deadly green powder on the floors and sweeping it into cracks and crevices, pouring the same powder under the paper which lined the silverware drawer and the shelves of the kitchen cabinet, or as she pulled apart the beds and poured an equally deadly liquid along the edges of the mattresses to discourage bedbugs, she reminded herself that being poor was no

SOURCE: David E. Bumbaugh, "The Forgotten Children of Affluent America," http://www.uc.summit.nj.uua.org/Sermons/DEB/980495.html. Published by permission of the author.

excuse not to be clean and free of bugs. And as we listened to the frequent sounds of ambulances and police cars screaming through the neighborhood in response to drunken brawls and family violence, she admonished us that just because we were poor was no reason not to be proud of ourselves and careful of our behavior.

She and my uncle worked hard to create an island of calm and peace and tranquillity in the midst of a world which was chaotic and unpredictable and sometimes dangerous. And like many poor parents, they invested their lives and their hopes in their children, in a future which would be qualitatively better than the lives they had known. They used to joke with us about that far-off day when one of us would be living in the White House, and how proud they would feel when we invited them to visit. We all knew that it would never happen, but we also heard in that standing joke the expectation that our lives would be better than theirs had been.

Like most poor parents, they had learned that life was never going to be much better for them and so they had invested their hopes for the future in their children. And like most poor parents, they simply did not know how to make that hope come true. They did what they could to protect us from the environing hazards of our world. They could be fierce when they perceived an unjust attack upon us. At the same time, they felt cowed and impotent in the face of authority and they did not know how to use the system to transform dreams into reality. They were often reduced to futile rage against a society which treated them as ciphers, which failed to honor their dreams, which saw their children not as promise or opportunity but as burden and expense.

We boys made our way through a public school system which regularly telegraphed to us the message that not very much was expected of us and then out into a world in which neither of us felt welcome, or at home. And of the two of us, one escaped that world of our childhood and one did not. And society tried to convince us—and itself—that the difference was the result of merit. Both of us knew, and still understand, that it was more a matter of luck which of us would escape and which would become part of that forgotten America. I tell you this so you understand my prejudice, and where I come from; for I have never forgotten whence I have come or my responsibility to speak for the forgotten children of America.

I was born in the midst of the Great Depression. But the world I experienced was not shaped by that economic catastrophe; the nations' financial collapse only served to make forgotten America more visible. Nor did that world end with the Depression. Rosemary L. Bray, who was born more than 20 years later, describes very much the same world in her memoir, *Unafraid of the Dark*. Rosemary's world was the urban ghetto of Chicago and her poverty was complicated by racism and sexism. But the family she describes, its relation to the larger world as it struggled to survive on the fringes of the mainstream economy, is very similar to the milieu in which I grew up. Not much changed in those 20 years.

Rosemary's family, however, benefited from a welfare system which, while never generous, never intended to do more than keep people living on the thin edge of poverty, did provide the center piece for a patchwork of economic support which, when supplemented by the occasional, low-paying, unpredictable job and the sustaining structures of family and church, made it possible to raise a family of children, to give them a minimal level of health care, to keep them in school, and to provide them with a full-time mother who was present to protect them from the dangers, temptations and seductions of the outside world.

Rosemary's family had a greater success rate than mine—all of the children made it out of the culture of poverty in which they grew up. Rosemary Bray, herself an affirmative action baby, went from the Chicago ghetto to Yale and then on to a career in journalism, and is now preparing for the Unitarian Universalist ministry. Her siblings hold responsible jobs and lead solidly middle-class lives. After her children were grown, Rosemary's mother was able to leave the welfare system, as did most recipients of welfare. Reflecting on her own journey, Rosemary says:

> We thought it would be enough to get an education and a job, to marry and start a family, to pay our taxes and vote, be ordinary and unexceptional citizens. It now appears that we were wrong. In attempting to downplay the circumstances of our early lives we left others—mothers and children such as we once were—at the mercy of ignorant and vicious ideologues who have never regarded the poor with anything but contempt. Changes in the welfare system since the late 1980s have made it nearly impossible for this story to happen today.

Reflecting on my own experience, and on the engrossing story my friend Rosemary Bray relates, I understand the despair I feel when I think about the plight of the poor in this vastly affluent nation, the even deeper despair when I think about the children who have been abandoned, who have been forgotten, who have fallen through the safety net as we pursue with determination a social policy shaped and driven by free-market values rather than human values.

There is a scandal at the heart of our national life. . . . In this nation, . . . we choose not to remember our fellow citizens who are not riding this wave of economic growth; we choose not to remember that our unprecedented prosperity is occurring in the context of the greatest disparity between rich and poor in the recent history of this nation; we choose not to remember that social policy has been structured so as to deprive the poor of the crumbs which once were their right as entitlements and to encourage the continuing concentration of resources in fewer and fewer hands; we choose not to remember that we have evolved the best political system money can buy—one which deprives the poor of any effective voice. As the Dow zig-zags . . . and we calculate what it will mean for the small part of it we own in retirement funds or investments, we choose not to remember that children under 6 years of age comprise the poorest age group in this nation and that our social policies are designed to beggar them even more.

There is a scandal at the heart of our national life. . . . It has to do with the fact that, finding ourselves in the midst of the hottest housing market in decades, busily buying and selling and building houses as if they were commodities, no longer worried about having to pay capital gains taxes on our profits, we choose not to remember in the midst of this frenzy that there are over 2 million homeless children in the United States. We choose not to remember in the midst of this frenzy that the average age of a homeless person in the United States is 9 years of age—the average age. While we congratulate ourselves on having the most dynamic economy in the world, we choose not to remember that we have failed to house our poorest citizens.

Statistics are telling and not all of them are moving up. In the wake of the program "to end welfare as we know it," we have determined that everyone must have a job and that there must be a lifetime limit for any kind of public assistance. But we have failed to insist that in exchange for full-time work, a person should receive a living wage. For a single mother with a preschool child in New Jersey, the cost of housing and day care are estimated to be $725 a month. If that mother can find a full-time job paying $5.10 an hour, she will take home about $700 a month. She will be in deficit before she begins to consider the cost of food, clothing, health care, and transportation.

And what is true in New Jersey is true in most of the nation as we who have never known welfare congratulate ourselves that we have "ended welfare as we know it." What we have done is to reincarnate the old work houses of Dickens' England. Only now they are low-paying jobs with no future, with no benefits, and no hope of ever moving out of poverty. At the same time, this policy separates children from their mothers' care for much of their young lives. The underlying effect of the ending of welfare, no matter how we dress it up in the moral language of the dignity of work, is to secure a steady source of low-paid workers for the economy. In their book, *The Breaking of the American Social Compact,* Piven and Cloward quote from Lawrence Mead's *Beyond Entitlement* to demonstrate the forces driving the recent assault on the poor. They quote Mead as saying in urging the attack on welfare:

> Low-wage work apparently must be mandated just as the draft has sometimes apparently been necessary to staff the military. Authority achieves compliance more efficiently than benefits. Government need not make the desired behavior worthwhile to people. It simply threatens punishment (in this case, the loss of benefits) if they do not comply.

And, of course, forcing people to accept jobs which fail to pay a living wage does not guarantee that such jobs will always be available for the unskilled and inexperienced. Statistics recently released suggest that across the nation only about 30 percent of those forced off welfare have been successful in finding jobs, and that statistic counts as successful anyone who has earned at least $100 in a three-month period. It is small wonder that food banks all over the country are reporting a 17 percent increase in requests for emergency food aid. And it is telling that almost two-fifths of those receiving emergency food aid are 18 or under, that nearly one-half of the families receiving such aid from food banks have children under 5. And these statistics are from the early part of 1997, before some of the largest cuts in food stamps and other resources for the poor took effect. In the midst of the nation's dynamic and vibrant economy, America's forgotten children are going hungry.

Poor people, by and large, have accepted the reality that they probably will not escape the trap of poverty, but they continue to hope that with a little luck, their children might. Even the most poverty-stricken ghettos are not devoid of dreams for the future. In his book *There Are No Children Here,* Alex Kotlowitz chronicles the lives of two small boys growing up in a Chicago housing project. Over and over again, like a small candle flickering in the overpowering gloom of poverty and despair, one hears in the voices of the boys and of their mother the inchoate hope that the children somehow will escape. Most of them do not. Once more, the fierce disparities between the rich and the poor drain that hope of any reality.

In 1991, Jonathan Kozol explained in his book *Savage Inequalities* part of the reason that the flickering hope for the children of the poor so often dies. He examined the vast gulf between the schooling of the children of the rich and the children of the poor. Poor children are concentrated in systems which spend significantly less per child upon education, and in districts where a significantly larger percentage of the school budget must be spent upon maintaining aging and derelict buildings. Poor children have fewer books and resources, less well-equipped libraries and labs, less experienced teachers, and larger class sizes than their affluent contemporaries. We in New Jersey know about this, as our courts struggle against entrenched political realities to enforce some kind of equality of resources on the state schools. Most of us would be quite willing to have equality in schooling, provided it did not affect the quality of our schools. In defense, we often insist that money is not the key to education, that if those other districts were better run, they would get more return on their investment, all the while fiercely fighting any effort to reduce the money spent by our schools, for fear that it would have a negative impact upon the quality of our children's education. And year after year the inequality grows and the children graduate from their unequal schools, some to promise and some to despair.

There is a scandal at the heart of our national life. . . . The scandal has to do with the fact that . . . two societies have been created—one of affluence and one of poverty, one of hope and one of despair, one fully in command of the national birthright and one condemned to be forever outside the charmed circle of the American dream. The scandal has to do with our determined blindness to the fact that what is at stake in our policies toward the poor, toward affluent America's forgotten children, is nothing less than the soul of the nation. We cannot continue a policy which subordinates human beings to market values without destroying our own humanity. A people is judged by how it treats its poorest, weakest, least advantaged members.

In his book *Waiting for the Barbarians,* Lewis Lapham remarked:

> The barbarian implicit in the restless energies of big-time, global capitalism requires some sort of check or balance, if not by a spiritual doctrine or impulse, then by a lively interest in (or practice of) democratic government. The collapse of communism at the end of the Cold War removed from the world's political stage the last pretense of a principled opposition to the rule of money, and the pages of history suggest that oligarchies unhindered by conscience or common sense seldom take much interest in the cause of civil liberty.

What Lapham remarks about the global condition is equally true of our national condition. Our children—and they are all our children, regardless of the political jurisdiction in which they live or the school districts in which they are educated—lay upon us a moral responsibility. We cannot be a moral people and permit our children to go hungry and homeless while the rich grow richer; we cannot be a moral people and educate some of our children for success while grooming others to provide the forced labor at the bottom of the labor market or starve; we cannot be a moral people if we accept luxury as our right and forget those, especially the children, who are condemned to lives of despair by the very policies which benefit us. We cannot be a moral people if we do not demand social policies which seek to ameliorate the discrepancies between the rich and the poor, policies which seek to provide the kind of support which allowed Rosemary Bray and countless others to emerge from poverty and to make of their talents a gift to the nation and the world.

"It is no disgrace to be poor," said my Aunt Martha. And she was right. But it is a disgrace to be affluent in America and to forget those others, who were children when we were children, who are contemporaries of our children and our grandchildren, who are not so blessed. It is more than a disgrace. It is a sin to turn our backs on the forgotten children of affluent America.

REFLECTION QUESTIONS FOR THE BOOK

1. Make a list of the major points from the two theoretical perspectives. Overall, do you find more support for one theoretical perspective over another? Find at least one selection from each chapter to justify and support your answer.
2. At this point, you should fully understand the sociological imagination and be able to apply it to all of the readings in the book, as well as to your own life. Take a moment to analyze your own life situation. How have you been affected by social structures and forces?
3. What social policies would the two theoretical perspectives advocate for solving the problem of poverty?
4. Reflect on the essays in Part III. How do poor people adapt to and cope with their difficult situations? How are their experiences structured by race, class, gender, age, and location?
5. Consider the selections in Part IV. How are the life chances of poor people affected by society's institutions?
6. Regarding agency, will positive social change occur if it is left up to those in power? If not, what is the best course for social change?
7. All of the selections in this book are meant to teach us something about poverty from the perspectives of those experiencing poverty. Make a list of lessons from Parts II through V and state the policy implications of each lesson.

Glossary

agency *See* human agency.

alienation The separation of human beings from each other, from themselves, and from the products they create.

ascribed characteristics Personal characteristics inherent in individuals over which they have no control such as age, family of origin, race, and sex.

assimilation The process by which individuals or groups voluntarily or involuntarily adopt the culture of another group, losing their original identity.

caste A stratum in the stratification system that is fixed at birth and is permanent.

culture of poverty The view that the poor are qualitatively different in values and lifestyles from the rest of society and that these cultural differences explain continued poverty.

cultural genocide The deliberate and systematic extermination of a national or racial group.

empowerment The feeling of control over outcomes.

environmental racism/classism The concentration of poor and racial minorities near hazardous and toxic environments.

fatalism The acceptance of all things and events as inevitable.

gentrification The redevelopment of poor and working-class urban neighborhoods into middle- and upper-middle class enclaves; often involves displacement of original residents.

grassroots organizations Collective efforts by common people to produce social change in their interests.

hierarchy The arrangement of people or objects in order of importance.

human agency Individuals are not passive. They actively shape social life by adapting to, negotiating with, and changing social structures.

institutional racism Occurs when the social arrangements and accepted ways of doing things in society disadvantage a racial group.

life chances The chances throughout one's life cycle to experience the good things of life.

living wage A wage that places the worker above the poverty threshold.

Medicaid The government health program for the poor.

Medicare The government program that provides partial coverage of medical costs for the elderly.

meritocracy A system of stratification in which rank is based purely on achievement.

near-poor People with family incomes at or above their poverty threshold but below 125 percent of their threshold.

poverty A standard of living below the minimum needed for the maintenance of an adequate diet, health, and shelter.

poverty threshold The poverty line established by the government based on the minimal amount of money required for a subsistence level of life. This threshold varies with the size of the family.

progressive tax An income tax where the rates increase with income.

race A group socially defined on the basis of a presumed common genetic heritage resulting in distinguishing physical characteristics.

regressive tax A tax rate that remains the same for all people rich and poor. The result is that poor people pay a larger proportion of their earned income than do affluent people.

severely poor People living at or below half of their poverty threshold.

social class A number of people who occupy the same relative economic rank in the stratification system.

social Darwinism The belief that the principle of the survival of the fittest applies to human societies, especially the system of stratification.

social determinism The assumption that human behavior is explained exclusively by social factors.

social mobility Movement by an individual from one social class to another.

social movement A collective attempt to promote or resist change.

social stratification Occurs when people are ranked in a hierarchy that differentiates them as superior or inferior.

socioeconomic status The measure of social status that takes into account several prestige factors such as income, education, and occupation.

stereotype An exaggerated generalization about some social category.

stigma A label of social disgrace.

structural approach to poverty The understanding of poverty resulting from factors external to individuals.

subculture A relatively cohesive cultural system that varies in form and substance from the dominant culture.

sweatshop A substandard work environment where workers are paid less than the minimum wage and where other laws are violated.

underclass The term used by culture of poverty theorists to denote the 10 to 20 percent of the poverty population that is mired in poverty with a strong propensity for crime and other antisocial behaviors, welfare dependency, and out-of-wedlock births.

underemployment Being employed at a job below one's level of training and expertise.

undocumented immigrants Immigrants who have entered the United States illegally.

upward social mobility Movement from a social class to a higher class (e.g., from poverty to working class).

welfare Receipt of financial aid and/or services from the government.

workfare The imposition of compulsory work programs and other mandatory requirements on welfare recipients with the goal of maximizing work participation while reducing reliance on welfare.

Web Sites

POVERTY: GENERAL

Center on Budget and Policy Priorities <*http://www.cbpp.org*> Articles and information on economic issues, including poverty and welfare.

Children's Defense Fund <*http://www.childrensdefense.org*> A children's advocacy group that provides data and analysis of trends concerning the plight of children.

Economic Policy Institute <*http://www.epinet.org*> Brief articles and data on various economic problems. <*http://www.inequality.org*> A clearinghouse for information on social class issues.

Joint Center for Poverty Research seeks to advance our understanding of what it means to be poor in America < *http://www.jcpr.org*>

Population Reference Bureau and Social Science Data Analysis Network <*http://www.AmeriStat.org*> A data bank on the most recent statistics, including poverty.

U.S. Bureau of the Census. Official data on poverty, published annually. <*http://www.census.gov/hhes/www/poverty.html*>

Worldbank provides resources and support for people and organizations working to understand and alleviate poverty <*http://www.worldbank.org/poverty*>

TOPIC AREAS

Activism

Center for Law and Social Policy <*http://www.clasp.org*>

Homeless advocates newspaper <*http://www.bigissue.com*>

National Law Center on Homelessness and Poverty <*http://www.nlchp.org/*>

Sticky Wicket: Poverty's Home Page <*http://www2.ari.net/home/poverty/*>

Welfare Law Center: <*http://www.welfarelaw.org*>

Welfare Rights Organizing Coalition <*http://www.scn.org/activism/wroc/*>

Working for Equality and Economic Liberation (WEEL) <*http://www.marsweb.com/~weel*>

Health

Kaiser Family Foundation <*http://www.kff.org/*>

National Center for Health Statistics <*http://www.cdc.gov/nchswww/index.htm*> The government's clearinghouse for health information, including health statistics on the poor.

U.S. Census Bureau facts about children's health insurance <*http://www.census.gov/hhes/hlthins/chldhins.html*>

U.S. Census Bureau statistics on health insurance coverage <*http://www.census.gov/hhes/www/hlthins.html*>

U.S. Department of Health and Human Services Administration on Aging <*http://www.aoa.dhhs.gov*>

Housing/Homelessness/ Neighborhoods

National Coalition for the Homeless <*http://www.nationalhomeless.org*>

National Law Center on Homelessness and Poverty <*http://www.nationalhomeless.org*>

U.S. Census Bureau: housing and household economics statistics <*http://www.census.gov/ftp/pub/hhes/www*>

Parenting/Children

Action Alliance for Children: Current trends and policy issues affecting children and their families in California <*http://www.4children.org*>

Administration for Children and Families < *http://www.acf.dhhs.gov*>

Center on Fathers, Families, and Public Policy—focus on never-married, low income fathers and their families <*http://www.cffpp.org*>

Center for Research on Child Wellbeing, Princeton University < *http://crcw.princeton.edu*>

Children's Defense Fund *http://www.childrensdefense.org*>

National Center for Children in Poverty <*http://cpmcnet.columbia.edu/dept/nccp*>

Resources for organizations to strengthen their programs and policies for children and families. <*http://www.handsnet.org*>

Racism/Discrimination

Administration for Native Americans—goal of social and economic self-sufficiency <*http://www.acf.dhhs.gov/programs/ana*>

Center for Immigration Studies, publications and policy <*http://www.cis.org*>

Institute on Race and Poverty, University of Minnesota <*http://www1.umn.edu/irp*>

Poverty, Race and Inequality Program, Institute for Policy Research, Northwestern Univ. <*http://www.nwu.edu/IPR/research/respoverty.html*>

Schools/Schooling

National Institute for Urban School Improvement <*http://www2.edc.org/urban/*>

U.S. Dept. of Education, National Institute on the Education of At-Risk Students <*http://www.ed.gov/offices/OERI/At-Risk*>

Urban Education Web/ERIC <*http://eric-web.tc.columbia.edu*>

Survival and Finances

America's Second Harvest <*http://www.secondharvest.org*> This website provides information on hunger in the U.S. and related public policies.

Brown University. Information regarding causes of and solutions to hunger. <*http://www.brown.edu/Departments/World_Hunger_Program*>

U.S. Census Bureau Income Statistics <*http://www.census.gov/hhes/www.income.html*>

Welfare

American Public Human Services Association, articles and information on welfare reform <*http://www.aphsa.org*

Applied Research Center <*http://www.arc.org*> Provides data on welfare reform.

Evaluations of Welfare to Work programs. <*http://www.welfareinfo.org/welftowork.htm*>

Making Wages Work: reports on programs that supplement income and wages in order to help families escape poverty and welfare dependency. <*http://www.makingwageswork.org*>

Welfare and Families <*http://epn.org/idea/welfare.html*>

Welfare Information Network <*http://www.welfareinfo.org*>

Work

Bureau of Labor Statistics <*http://stats.bls.gov/blshome.html*>

Environmental Justice Resource Center <*http://www.ejrc.cau.edu/*>

Rockefeller Foundation: publication on employment issues. <*http://www.rockfound.org*>

U.S. Department of Labor: Employment and Training Administration (ETA). ETA Welfare to Work <*http://wtw.doleta.gov/*>

Index

K

L

M

N

O

P